T0326454

Bangladesh Confronts Climate Change

ANTHEM CLIMATE CHANGE AND POLICY SERIES

The **Anthem Climate Change and Policy Series** aims to compile the best integrative thinking on the nature of and relationships among the processes underlying climate and other closely related forms of global change. It also seeks to identify how science can inform the development of adaptation and mitigation strategies, and what policies might be developed to most effectively implement those strategies. Climate change, with its links to agriculture, water resources, human health, ecosystems and global security, is introducing challenges that are stretching humanity's capacity for adaptation and effective mitigative action. Physical changes induced by climate warming and directly affecting human needs include polar ice loss, with its associated alteration of weather patterns including the jet stream; mountain glacier losses with implications for freshwater supplies and sea level rise; changing meteorological patterns with implications for global food supplies; land and ocean ecosystems change including the alteration of ocean acidity, a reduction in biodiversity and the loss of coral reefs and important food species. Climate warming–induced change is further complicated by direct human activities, such as major land surface transformation; the over-harvesting of global fisheries; and the industrial pollution of air, land and water. Merely defining the links between the climate warming–induced drivers of change and their many potential impacts is often a daunting problem. This series is geared towards bringing the best scholarship to bear in serving the practical environmental policy and management community.

Series Editor

Brooke L. Hemming – United States Environmental Protection Agency, USA

Bangladesh Confronts Climate Change

Change

Keeping Our Heads above Water

Manoj Roy, Joseph Hanlon and David Hulme

ANTHEM PRESS

Anthem Press
An imprint of Wimbledon Publishing Company
www.anthempress.com

This edition first published in UK and USA 2016
by ANTHEM PRESS
75–76 Blackfriars Road, London SE1 8HA, UK
or PO Box 9779, London SW19 7ZG, UK
and
244 Madison Ave #116, New York, NY 10016, USA

British Library Cataloguing-in-Publication Data
A catalogue record for this book is available from the British Library.

Library of Congress Cataloging-in-Publication Data
A catalog record for this book has been requested.

ISBN-13: 978-1-78308-632-0 (Hbk)
ISBN-10: 1-78308-632-7 (Hbk)

ISBN-13: 978-1-78308-633-7 (Pbk)
ISBN-10: 1-78308-633-5 (Pbk)

This title is also available as an e-book.

CONTENTS

ILLUSTRATIONS

Boxes

Maps

ABBREVIATIONS, ACRONYMS AND BANGLADESHI TERMS

$	United States dollar
£	UK pound
ADB	Asian Development Bank
AL	Awami League
Aman	flooded rice crop, *B Aman* is broadcast or directly seeded and *T Aman* is transplanted
APA	National Action Plan on Adaptation, MEF 2005, superseded by BCCSAP
Aus	rain-fed rice
AWD	alternate wetting and drying irrigation system
Bangla, Bengali	language of Bangladesh
BARC	Bangladesh Agricultural Research Council
barsha floods	normal beneficial floods which renew the land
basti, bosti, bustee	slum
BAU	business as usual
BBS	Bangladesh Bureau of Statistics
BCAS	Bangladesh Centre for Advanced Studies, Dhaka
BCCRF	Bangladesh Climate Change Resilience Fund, a World Bank–managed fund
BCCSAP	Bangladesh Climate Change Strategy and Action Plan, MEF 2008
BCCTF	Bangladesh Climate Change Trust Fund, a government-managed fund
BDRCS	Bangladesh Red Crescent Society
beel	natural shallow lake which fills with water during monsoon
BGMEA	Bangladesh Garment Manufacturers and Exporters Association
BIDS	Bangladesh Institute of Development Studies, Dhaka
BIWTA	Bangladesh Inland Water Transport Authority
BMD	Bangladesh Meteorological Department
BNP	Bangladesh Nationalist Party
bonna floods	less frequent but more destructive floods
Boro	irrigated, transplanted rice
Borsakal	Bengali rainy season
Bosontokal	Bengali spring

BRAC	originally Bangladesh Rehabilitation Assistance Committee and now Bangladesh Rural Advancement Committee, a very large NGO and social business
BRRI	Bangladesh Rice Research Institute
BUET	Bangladesh University of Engineering and Technology
BWDB	Bangladesh Water Development Board
CBO	community-based organization
CEGIS	Centre for Environmental and Geographical Information Services, Dhaka
CH_4	methane
char	new island created from sediment
ClimUrb	Poverty and Climate Change in Urban Bangladesh Research Project, Manchester
CNG	three-wheel taxis using compressed natural gas as fuel
CO_2	carbon dioxide – important greenhouse gas
COP	annual Conferences of the Parties which are signatories of the UNFCCC
CPP	Cyclone Preparedness Programme
CSIRO	Commonwealth Scientific and Industrial Research Organisation, Australia
DAE	Bangladesh Department of Agricultural Extension
DCC	Dhaka City Corporation
DDM	Department of Disaster Management
DESCO	Dhaka Electricity Supply Company
DFID, DfID	UK Department for International Development
district	second-tier administration unit (there are 64)
division	largest government administration unit (there are 8)
DSK	Dushtha Shasthya Kendra, a Dhaka NGO
ESCAP	United Nations Economic and Social Commission for Asia and the Pacific
ESRC	UK Economic and Social Research Council
EU	European Union
FAO	UN Food and Agriculture Organization
FAP	Flood Action Plan
FFWC	Flood Forecasting and Warning Center, Bangladesh Water Development Board
G7	Group of seven large industrialized countries
G77	loose coalition of developing nations, originally with 77 members
GCF	Green Climate Fund, donor fund within the UNFCCC mechanism
GDP	Gross Domestic Product, the value produced within a country's borders
GNI	Gross National Income (previously called GNP, Gross National Product), the value produced by all citizens, including remittances
Grissokal	Bengali summer

ha	hectare (100 m × 100 m)
HBRI	Bangladesh House Building Research Institute
HDI	UNDP Human Development Index
Hemontokal	Bengali late autumn
HFCs	hydrofluorocarbons
hilsa	a kind of fish, important in the Bangladeshi cuisine
ICAI	UK Independent Commission for Aid Impact
ICCCAD	International Centre for Climate Change and Development, Dhaka
IEG	World Bank Independent Evaluation Group
IIED	International Institute for Environment and Development, London
IMF	International Monetary Fund
INGO	international non-government organization
IPCC	Intergovernmental Panel on Climate Change
IRRI	International Rice Research Institute
IVR	Interactive Voice Response
IWM	Institute of Water Modelling, Dhaka
jeel, jheel	pond formed in a natural depression
jhupiri (house)	a small, temporary house made of straw, leaves or plastic sheets
JICA	Japan International Cooperation Agency
kg	kilogramme
khal	canal
Kharif-1	March–June crop season
Kharif-2	July–October crop season
killa	livestock pen
KJDRP	Khulna-Jessore Drainage Rehabilitation Project
km	kilometre
kutcha (house)	a self-built house of clay or bamboo walls
LDCs	least developed countries
LIC	Low Income Community
LNG	Liquefied Natural Gas
m	metre
mastaan	local muscleman or enforcer
MDGs	Millennium Development Goals
mm	millimetres
MOEF	Ministry of Environment and Forests
monga	hungry season, just before harvest
$MtCO_2e$	Megatonnes CO_2-equivalent
N_2O	nitrous oxide
NGO	non-government organization
NSDF	National Slum Dwellers Federation, India
PFCs	perfluorocarbons
PKSF	Palli Karma-Sahayak Foundation, Bangladesh government
pucka (house)	a well-constructed cement or brick house
Rabi	November–February crop season

SF$_6$	sulphur hexafluoride
Sitkal	Bengali winter
Soratkal	Bengali pre-autumn
SPARC	Society for the Promotion of Area Resource Centers, India
SRI	System of Rice Intensification
SWC	Bangladesh Meteorological Department Storm Warning Centre
t	tonne
Taka, TK	Bangladesh currency; TK 69 = $1 before 2011, TK 78 = $1 after 2013
TIB	Transparency International Bangladesh
TRM	Tidal River Management
TVA	US Tennessee Valley Authority
UN	United Nations
UNDP	United Nations Development Programme
UNEP	United Nations Environment Programme
UNFCCC	United Nations Framework Convention on Climate Change
union parishad (UP)	lowest level government administration unit
upazila	sub-districts, previously called *thanas* (there are 489)
US	United States of America
USAID	United States Agency for International Development
USSR	Soviet Union
WAPDA	East Pakistan Water and Power Development Authority
WASA	Dhaka Water Supply and Sewerage Authority
WEF	World Economic Forum
WMO	World Meteorological Organization
WMOs	community water management organizations
zamindar	local ruler in the Mughal period and tax collector in the colonial period

ACKNOWLEDGEMENTS

It would not have been possible to write this book without the support and good humour of hundreds of villagers and shack-dwellers in Bangladesh who answered our questions and helped us understand how they manage the complex and difficult environments in which they live. Bangladeshi researchers, intellectuals, climate negotiators, government officials and journalists gave generously of their time and ideas. Colleagues in the UK at the universities of Lancaster and Manchester and at the International Institute for Environment and Development helped us work through data and findings. Particular thanks to Clive Agnew, M. Asaduzzaman, Niki Banks, Hugh Brammer, Sally Cawood, Nigel Clark, David Dodman, Simon Guy, Syed Hashemi, Abul Hossain, Saleemul Huq, Kairul Islam, Ferdous Jehan, Hafij Khan, Z. Karim, Anirudh Krishna, David Lewis, Golam Iftekhar Mahmud, Fuad Mallick, Diana Mitlin, Ainun Nishat, Afroza Parvin, Atiq Rahman, Hossain Zillur Rahman, James Rothwell, Mustafa Saroar, David Satterthwaite, Satchidananda Biswas Satu, Binayak Sen, Dibalok Singha, Teresa Smart and Khurshid Zabin Hossain Taufique.

The research that underpins this book has been supported by a number of sponsors. The initial idea that set this project in motion was funded by the Rory and Elizabeth Brooks Foundation back in 2008. Our deep thanks to the foundation for its patience in waiting so long for the main outputs. Further research and primary fieldwork costs were supported by the ESRC/DFID grant (RES-167-25-0510) 'Community and Institutional Responses to the Challenges Facing Poor Urban People in an Era of Global Warming in Bangladesh' under the ClimUrb (Poverty and Climate Change in Urban Bangladesh) project, and by the UKRC's Environmental Services for Poverty Alleviation (ESPA) grant (NE-L001616-1) 'Institutions for Urban Poor's Access to Ecosystem Services: A Comparison of Green and Water Structures in Urban Bangladesh and Tanzania' (EcoPoor). In addition, research for the project on 'Climate Change and Urban Vulnerability in Africa' (CLUVA) funded by the EU Framework Programme 7 (EU-FP7) also helped our thinking.

At the University of Manchester our work has benefited from advice and comments from colleagues at the Global Development Institute (uniting the Institute for Development Policy and Management and the Brooks World Poverty Institute). Without administrative and logistical support from Kat Bethell, Julia Brunt, Chris Jordan, Julie Rafferty and Denise Redston at the Global Development Institute it would not have been possible to complete this work. At the University of Lancaster our work has benefited from comments from colleagues at the Society and Environment research cluster within Lancaster Environment Centre and a Research Incentivisation Grant to partially cover our field costs.

ABOUT THE AUTHORS

Manoj Roy is lecturer in sustainability at Lancaster Environment Centre, Lancaster University, UK. He was born in Bangladesh and spent his childhood in rural Bangladesh. He received a Bachelor of Architecture degree from Bangladesh University of Engineering and Technology. He worked in Dhaka as an architect before moving to Germany and ultimately to UK for higher education and academic work. He has degrees in architecture, infrastructure planning and urban planning. He specializes in poverty analysis through interdisciplinary methods combining technical analysis (e.g., architectural and planning, spatial analysis and modelling) with a social (e.g., livelihoods, well-being) and political (governance, institutional) analysis. Working with David Hulme, he co-directed the ClimUrb (Poverty and Climate Change in Urban Bangladesh) project, and is currently the principal investigator of the EcoPoor (Institutions for Urban Poor's Access to Ecosystem Services: A Comparison of Green and Water Structures in Bangladesh and Tanzania) and PSlums (The Rise of Private Slum Dwellers in Bangladesh and India: Heroes or Villains?) projects. His research sponsors include British Academy, ESRC-DFID Joint Scheme of Research on International Development, RCUK's Ecosystem Services for Poverty Alleviation (ESPA) and EU FP7. His research countries include Bangladesh, India and Tanzania.

Joseph Hanlon is visiting senior fellow at the Department of International Development of the London School of Economics and visiting senior research fellow at the Open University, Milton Keynes, UK. He is a journalist and author or editor of more than a dozen books, looking at how the international context can be changed to give people in the Global South more power over their own development strategies. His most recent books are *Galinhas e cerveja: Uma receita para o crescimento* (*Chickens and beer: A recipe for agricultural growth in Mozambique*, with Teresa Smart, 2014), *Zimbabwe Takes Back Its Land* (with Teresa Smart and Jeanette Manjengwa, 2013), *Just Give Money to the Poor* (with David Hulme and Armando Barrientos, 2010), *Do Bicycles Equal Development in Mozambique?* (2008) and *Civil War, Civil Peace* (with Helen Yanacopulos, 2006). He is editor of the *Mozambique Political Process Bulletin* and was policy officer for the Jubilee 2000 campaign to cancel the unpayable debt of poor countries.

David Hulme is professor of development studies and executive director of the Global Development Institute at the University of Manchester, UK. He has been researching and writing on poverty and well-being in Bangladesh for almost thirty years. His recent books include *Should Rich Nations Help the Poor?* (2016), *Urban Poverty and Climate Change* (with Manoj Roy, M. Hordijk and S. Cawood, 2016), *Global Poverty* (2015) and *Social Protection in Bangladesh* (with H. Z. Rahman, M. Maitrot and P. Ragno, 2014).

Chapter One

ACTORS, NOT VICTIMS

Sumi is busy raising the floor of her small but comfortable house in Rupshaghat, Khulna, to make sure that her bed and TV set are not damaged when the area floods during the next monsoon. Khondaker Kabir is designing houses that local carpenters can build and also better withstand the stronger cyclones. Ainun Nishat was vice chancellor of BRAC University and led the Bangladesh negotiators at United Nations climate change conferences where a mixed government and civil society team won concessions from the developed world on payments for loss and damage caused by climate change. All three are Bangladeshis on the front line responding to climate change. They are not 'victims', but people who know climate change is real and that Bangladesh must adapt. And Bangladeshis have centuries of experience in adapting to environmental challenges.

Bangladesh is the eighth most populous country in the world but only slightly larger than England and the same size as the US state of Illinois. It is the most densely populated country in the world – with double the population density of Taiwan and triple that of the Netherlands and Rwanda.[1] It has so many people because it lies on a rich delta and feeds itself. But that richness comes at a price. One of the largest deltas in the world, it is mainly flat and low lying. Water from the Himalayas and monsoon rains pours down the Ganges and Brahmaputra rivers, causing annual flooding. Some floods are benign and bring fertility, but others are hugely destructive and the rivers shift their courses eroding farmland and creating new islands. Cyclones coming north, up the Bay of Bengal, can cause massive damage. And the environment itself is hugely variable, from year to year and place to place. Bangladesh is only 600 km wide, but average annual rainfall ranges from 1.5 m in the west to 5 m in the east.

Indeed, Bangladesh has been and is shaped by geographic, environmental and political factors often entirely beyond its borders. Climate change is the newest of these factors. But centuries of coping, adapting and shaping the land and society have given Bangladeshis a strong understanding of what climate change will bring. Knowing how devastating unchecked climate change will be has propelled Bangladeshi academics and politicians into leading roles in international negotiations. National scientists and engineers, as well as community groups building on historic knowledge, are already working

---

1   Bangladesh had a population of 160 million at the end of 2015, compared to India with 1.3 billion and Pakistan with 192 million. Bangladesh's area is 148,000 km². Illinois is 150,000 km² and England is 130,000 km² (the entire UK is 240,000 km²). Bangladesh has the highest population density excluding small island and city states. Its population density is 1,100 per km², compared to Taiwan (650), Rwanda (440) and the Netherlands (410). (All 2015 official figures)

to protect Bangladesh against the vagaries of the existing climate as well as the problems that will be created by future climate change.

Thus Bangladeshis have centuries of adaptation experience, including ancient systems of dykes, siting hamlets on small hills and building houses on raised earthen plinths. The 45 years since independence have seen many local innovations including improved varieties of rice which have changed the entire cropping pattern. These are Bangladeshi innovations. Lack of understanding of local systems means advice and projects from British colonizers, the World Bank, the United States and the Netherlands have often been problematic.

The biggest success has been cyclone protection, which is little recognized outside Bangladesh because it is a national innovation and not one driven by donor agencies. There have been three 'super cyclones' in the past 50 years with wind speeds over 222 km/h: Bhola (1970), which left between 200, 000 and 500,000 dead; Gorky (1991), which left 138,000 dead; and Sidr (2007), which left 3,363 dead.[2] This huge drop in fatalities is due to national actions – better early warning, cyclone shelters and higher coastal dykes. The Bangladesh Meteorological Department follows all tropical depressions closely and issues accurate and detailed cyclone warnings. The Cyclone Preparedness Programme has 50,000 volunteers who use megaphones and house-to-house contact to give cyclone warnings. There are now more than three thousand and seven hundred cyclone shelters – huge concrete structures often used as schools that are built on pillars at first-floor level, so wind and water can pass under and around them. They are stocked with food and water and serve entire communities. Finally coastal embankments are being raised, which at least breaks the violence of the storm surge.

'Bangladesh is ranked as one of the most climate vulnerable countries in the world' according to the 2014 *Fifth Assessment Report* of the Intergovernmental Panel on Climate Change (IPCC).[3] Although it does not create any new problems for Bangladesh, climate change will accentuate the present problems. More variable and more severe weather makes all the adaptation challenges much more serious. More droughts, more floods, more erosion, more severe cyclones and more salt penetration require better rice varieties, higher embankments, stronger houses built on higher platforms and more cyclone shelters. Urban migration means adapting the country's burgeoning cities to deal with a worsening environment, which will be a major challenge.

But the people of Bangladesh are not simply victims – climate change refugees – as the Western media proclaims. Bangladeshis are taking a lead, both nationally and

---

2   Department of Disaster Management, *Disaster Report 2013* (Dhaka: Department of Disaster Management, Ministry of Disaster Management and Relief, 2014), 22, gives the wind speeds. Bay of Bengal cyclones are categorized under the India Meteorological Department system, according to which above 222 km/h sustained wind, a storm is a 'super cyclone' or category 5 cyclone. Category 4 is 167–221 km/h and called an 'extremely severe cyclone'. Accessed 24 April 2016, http://imdtvm.gov.in/index.php?option=com_content&task=view&id=15&Itemid=30.

3   Intergovernmental Panel on Climate Change (IPCC), 'The IPCC's Fifth Assessment Report: What's in It for South Asia' (London: Climate and Development Knowledge Network, 2014), 18.

internationally, in responding to climate change, and in this book we talk to them. As Prof. Nishat told us, 'Bangladesh is God's laboratory on natural disaster, so we know how to deal with most of this.' We hear from people living in informal settlements in cities trying to keep dry, and from negotiators trying to convince the already industrialized nations – and the rapidly growing ones as well – that they must reduce their carbon emissions.

Scientific research is central to this story, from the researchers modelling climate change to those developing the new rice varieties. They are working with local people, for example, to use traditional methods to raise the land level to stay ahead of sea level rise. If climate change is checked, Bangladesh can adapt, and keep its head above water. But who pays?

And adaptation can only go so far. Embankments can be raised to cope with sea level rise and pumps can be installed to deal with extreme rainfall, but there are limits. Unchecked climate change will drown parts of Bangladesh.

Bangladesh is one of the countries most vulnerable to climate change, which is not the result of Bangladeshi actions. Yet again, the people must respond to outside forces. Bangladeshis refuse to be victims – they will not sit passive and helpless in the face of a possible catastrophe. Instead negotiators fighting to curb rich countries' emissions, scientists producing better adapted rice and ordinary people raising their land are combating and adapting to climate change. They are on the front line of climate change and this book is their story.

Shaped by Geography and Politics

Bangladesh was carved first from colonial India and then from Pakistan, which explains its unusual borders. Bangladesh is surrounded on three sides by India, but the northernmost point of the country is just 25 km from Nepal and 65 km from Bhutan. (Indeed, at its nearest point, China is only 100 km north of Bangladesh.) In the southeast Bangladesh has a 271 km border with Myanmar (Burma). The south faces the Bay of Bengal.

Three vast rivers – Brahmaputra, Ganges and Meghna – join together in Bangladesh and make the third largest river in the world (by flow volume, after the Amazon and Congo rivers). All three receive most of their water from monsoon rain, and normally 90 per cent of the water comes from outside Bangladesh:[1]

- The Brahmaputra is 3,000 km long and provides 56 per cent of the total water. It flows from the northeast and receives 27 per cent of its water from snow and melting glaciers in the Himalaya mountains above 2,000 m in northeast India and China. Because of snow and glacier melt, it starts to rise early, in March, and reaches its peak levels in July and August.
- The Ganges is 2,600 km long and provides 25 per cent of the total water. It flows from the northwest and receives 10 per cent of its water from snow and melting glaciers in

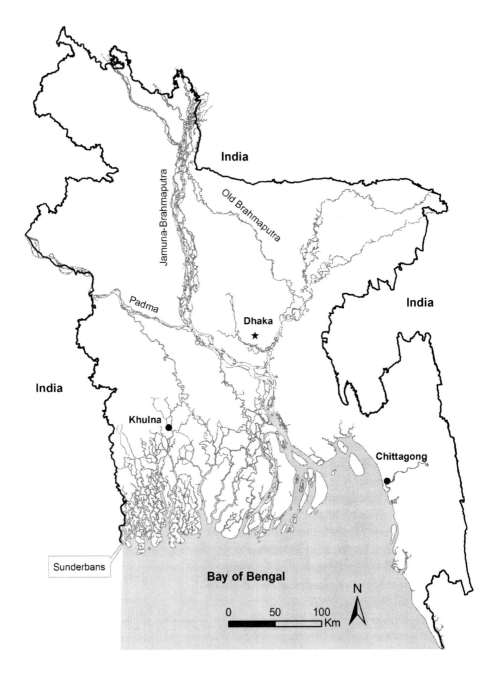

Map 1.1 Bangladesh, showing main cities and rivers, and its position in South Asia.
Map: Manoj Roy

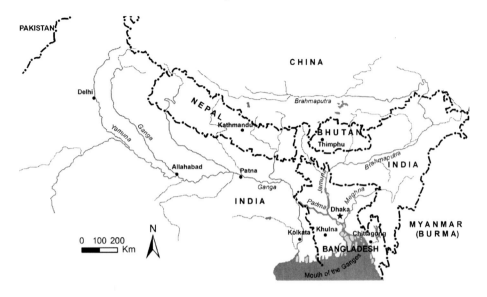

Map 1.2 The three main rivers flowing into Bangladesh rise in China and India.
Map: Manoj Roy

the Himalaya mountains above 2,000 m in northern India and Nepal. This river starts to rise in May and peaks in August and September, later than the Brahmaputra.
• The Meghna is 264 km long and provides 19 per cent of the total water. The wettest place in the world is in the state of Meghalaya, India, just 20 km north of the Bangladesh border, and it receives nearly 12 m of rain per year, and twice that in exceptional years, which flows in the Meghna. High water is July through September.

Floods are a natural part of the ecology of Bangladesh, and most floods are not 'disasters'. However three rivers with different catchment areas, each with its own rain and snowfall patterns, combined with Bangladesh's own variable rainfall, means every flood is different. Bangla has two words for floods, *barsha* for the normal beneficial floods which renew the land and *bonna* for the less frequent but more destructive floods that happen about once a decade.

Barsha floods typically inundate 20 per cent of the country's land for three to five months. But peak levels last only for a week or two during the monsoon; people move to higher ground and children see it as a bit of a party. One of the authors remembers as a child using a banana tree trunk as a makeshift boat and paddling around his farm – the flood season was 'fun'.

The worst flood in a century was in 1998 when 68 per cent of the country was flooded, including 70 per cent of the capital, Dhaka; and in many areas flooding lasted for two months. In that year, rainfall was unusually heavy and all three rivers hit their peaks at the same time, which coincided with a spring high tide that blocked water from flowing into the sea.[5] Yet the 1998 flood came only four years after the driest year, 1994, when a mere

---

5   Brammer, *Can Bangladesh Be Protected*, 124–28.

Sumi's self-built house in Rupshaghat slum in Khulna is small but comfortable. The slum floods each year and the floors of the houses are raised, but the flood in 2012 was higher than ever before, so she has raised her furniture – the sideboard and bed – on blocks to keep them above the new higher flood levels.

Photo by Joseph Hanlon

0.2 per cent of Bangladesh was flooded. There are also regional differences, with most floods affecting just part of the country. Figure 1.1 shows the percentage of Bangladesh that has been flooded each year from 1960, and demonstrates the extreme variation.[6]

From the south come cyclones. Typically two to seven cyclones form in the Bay of Bengal each year, in the period just before and just after the monsoon. Of these, most do not reach land or make landfall in India or Myanmar, but, on average, one cyclone each year reaches land in Bangladesh and there is a severe cyclone every two to four years.[7]

---

6   Department of Disaster Management, *Disaster Report 2013*, 13–14, based on Bangladesh Water Development Board data.

7   For the 54-year period 1960–2013 inclusive, Hugh Brammer, *Bangladesh: Landscapes, Soil Fertility and Climate Change* (Dhaka: University Press, 2016), 144. Table 8.15 reports there were 38

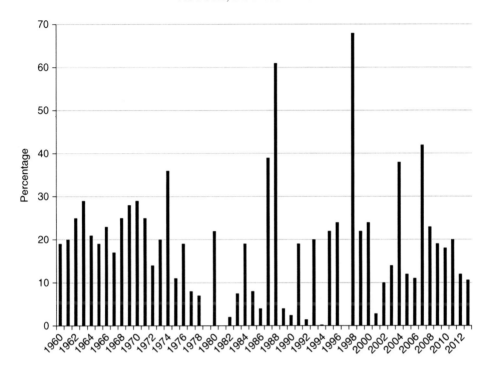

Figure 1.1 Maximum percentage of Bangladesh flooded each year
Source: Department of Disaster Management, *Disaster Report 2013* (Dhaka: Department of Disaster Management, Ministry of Disaster Management and Relief, 2014), 13–14, based on Bangladesh Water Development Board data. Data for 1979, 1981 and 1997 are not available and those years are omitted.

Bangladesh as 'Basket Case'

Just as there has been a constant struggle to deal with storms and floods coming from the outside, so there has been a battle against occupiers and foreign powers. We will only give the briefest summary of Bangladesh's history here as it relates to its environment, and point readers to David Lewis's excellent *Bangladesh: Politics, Economy and Civil Society* for more detail.[8]

Written records go back to the third century BC and there have been various occupiers and periods of local control. Bengali (or Bangla) is the language of Bangladesh and in the Indian states of West Bengal, Tripura and Assam. Until 1947 this was seen as a unified area and from the thirteenth century there were three main religions, Shia Islam, Sunni Islam and Hinduism. Despite substantial local resistance, the Mughal Empire gained control over Bengal at the end of the sixteenth century, and the region

---

cyclones, of which 23 were severe, while the Department of Disaster Management, *Disaster Report 2013*, 22, for the same period reports 45 cyclones of which 14 were rated 'severe'. There have been relatively fewer cyclones since 2000.

8   David Lewis, *Bangladesh: Politics, Economy and Civil Society* (Cambridge, UK: Cambridge University Press, 2011).

developed and prospered. Mughal power declined from 1707 and the British East India Company grew in power. But there was resistance to the company rule and in 1756 the British were defeated in Calcutta (now Kolkata). The city was recaptured in 1757 and in 1765 formal British government rule commenced. Sporadic resistance continued throughout India and British colonialism finally ended in 1947. India was partitioned, with two largely Muslim areas defined as a new country: Pakistan. Millions of Muslims fled to Pakistan and Hindus fled to India. But what is now Bangladesh retained a large Hindu minority. The largely Muslim country was split into two parts, East and West Pakistan, separated by 1,600 km. This was an unprecedented way to create a country and never worked. The Bengali East was effectively ruled by the Urdu-speaking West, which proved unacceptable and led to the growth of a new nationalist Bangla-speaking movement.

Meanwhile, the cold war between the capitalist United States (US) and socialist Soviet Union (USSR) began to play a role. In 1958 Gen. Ayub Khan staged a coup and established a military government in Pakistan that continued under Gen. Yahya Khan in 1969. With India aligned to the USSR, Pakistan sided with the US and China in the cold war. This led to substantial US and World Bank support for the Ayub Khan government. In 1970 the Bangladeshi Awami League won a majority in the Pakistan parliament but was blocked by the military from taking office. A few weeks later, on 12 November 1970, a cyclone and 5 m high storm surge killed between 200,000 and 500,000 people in the Chittagong area. The humanitarian relief was badly managed. The Pakistan government was blamed for the lack of preparation and a slow response, which intensified the pressure for Bangladesh's independence.

Operation Searchlight was organized at the highest level of the Pakistani military and launched unexpectedly on 25 March 1971. West Pakistani soldiers in East Pakistan began a systematic massacre of civilians who were believed to be promoting independence. Hindus and intellectuals were particular targets; tanks shelled the Dhaka University campus and students were machine-gunned as they fled. An official of the US consulate in Dhaka estimated at the time that at least 500 students were killed in the first two days.[9] Bangladeshi independence was declared on 26 March, but West Pakistan fought for a further nine months trying to retain control.

The US strongly backed its cold war allies in West Pakistan. Two weeks into the slaughter, Archer Blood, the US Consul General in Dhaka, and most of the consulate staff sent a telegram to Washington protesting against continued US support for West Pakistan and its use of US weapons to massacre innocent civilians. But Blood did not know that Ayub Khan was using his links with China to work with President Richard Nixon and National Security Advisor Henry Kissinger to set up their historic meetings in China.[10] The US backed West Pakistan throughout the war despite the atrocities it perpetrated.

---

9 Gary J. Bass, *The Blood Telegram: Nixon, Kissinger and the Forgotten Genocide* (New York: Alfred A Knopf, 2013) and Lewis, *Bangladesh*, 71.

10 Bass, *The Blood Telegram*.

As the war continued, India backed Bangladesh and the USSR backed India and Bangladesh; both the US and USSR sent warships to the Bay of Bengal. The nine-month war ended with the surrender of Pakistan's military to the Bangladesh-India Allied Forces on 16 December 1971. Between one and three million Bangladeshis had been killed, including a significant portion of the educated elite, who had been targeted by Pakistan.[11] Even now, 45 years later, it is notable that many of the key intellectuals working on climate change who were students at the time of the massacre have only survived because they were abroad at the time working on their PhDs.

Not surprisingly, the new nation and the US were hostile to each other. Just before independence, a US under secretary of state said Bangladesh would be 'an international basket case', to which Kissinger, pointing to his unwillingness to support the new country, replied, 'but not necessarily our basket case'.[12] There was a military coup in Bangladesh in 1975, and General Ziaur Rahman (known as Zia) eventually gained power. A former member of the Pakistani army who had been decorated for bravery in the 1965 Pakistani war against India and played a major role in Bangladesh's war of independence, Zia was pro-Western and won US support.

This sort of politics may seem very far from climate change and ecology, but just as a devastating cyclone played a key role in pushing forward independence, so the Mughals, British and cold war US shaped the dykes and canals that do and will play an important part in the response to climate change.

Bangladesh as a Development Success Story

Since independence Bangladesh has transformed dramatically, especially in comparison to neighbour India and former ruler Pakistan. Jean Drèze and Amartya Sen in their book *An Uncertain Glory: India and Its Contradictions*[13] point out that 'over the last two decades India has expanded its lead over Bangladesh in terms of average income (it is now about twice as rich in income per capita as Bangladesh), and yet in terms of many typical indicators of living standards (other than income per head), Bangladesh not only does better than India, but also has a considerable lead over it (just as India had, two decades ago, a substantial lead over Bangladesh in the same indicators).'

They go on to argue that 'there is much evidence to suggest that Bangladesh's rapid progress in living standards has been greatly helped by the agency of women, and par-ticularly the fact that girls have been rapidly educated and women have been widely

---

11 Lewis, *Bangladesh*, 71, 77.

12 The phrase 'international basket case' was used by Under Secretary of State for Political Affairs Ural Alexis Johnson at an internal interdepartmental meeting in the White House on 6 December 1971, just before Bangladesh independence. But it was soon leaked to the US press by some in the US government who had backed Pakistan in the war and were opposed to recognition of independent Bangladesh. Mohammad Rezaul Bari, 'The Basket Case', *Forum*, March 2008 (Dhaka: *The Daily Star*). Accessed 6 September 2016, http://archive.thedailystar.net/forum/2008/march/basket.htm.

13 Jean Drèze and Amartya Sen, *An Uncertain Glory: India and Its Contradictions* (Princeton: Princeton University Press, 2013).

Table 1.1 Human development details for Bangladesh and its neighbours

	Bangladesh	India	Pakistan	Myanmar	World
HDI 2015 rank (of 188)	142	130	147	148	
GNI per capita rank	147	126	133	136	
GNI rank – HDI rank	5	−4	−14	−12	
GNI per capita ($PPP)	3,191	5,497	4,866	4,608	14,301
Gender related					
Sex ratio at birth, male to female	1.05	1.11	1.09	1.03	1.05
Fertility rate – births per woman	2.2	2.5	3.2	2.0	2.5
Life expectancy at birth, years					
Female	72.9	69.5	67.2	68.0	73.7
Male	70.4	66.6	65.3	63.9	69.5
Expected years of schooling					
Female	10.3	11.3	7.0	...	12.2
Male	9.7	11.8	8.5	...	12.4
Maternal mortality (deaths/1000 live births)	1.7	1.9	1.7	2.0	2.1

Source: UNDP *Human Development Report 2015.*

All data from 2014 or most recent year.

The Human Development Index (HDI) is calculated each year by the United Nations Development Programme (UNDP) based on income, education and health.

Gross Domestic Product (GDP) is the value produced within a country's borders, whereas the Gross National Income (GNI, previously called GNP, Gross National Product) is the value produced by all the citizens. These are largely the same, except that remittances are included in GNI and not GDP. For Bangladesh this make a major difference because remittances are approximately 8 per cent of GNI.

involved – much more than in India – in the expansion of basic education, health care, family planning and other public services as well as being a bigger part of the industrial labour force.' Table 1.1 gives comparisons of Bangladesh with neighbours India and Myanmar and with Pakistan, and shows it is doing well on social indicators despite its lower income.

'Bangladesh has managed to sustain a surprisingly rapid reduction in the rate of child malnutrition for at least two decades [and] from 1997 to 2007 Bangladesh recorded one of the fastest prolonged reductions in child underweight and stunting prevalence in recorded history', according to a 2015 article in the prestigious journal *World Development*.[14] This is linked to the country's 'remarkable progress in poverty reduction and delivering effective health and family planning services'. There are more girls than boys in school and they finish more years of schooling, in part because of a secondary school stipend for girls.[15] After independence, when female employment was still uncommon, Bangladesh

14 Derek Headey, 'The Other Asian Enigma: Explaining the Rapid Reduction of Undernutrition in Bangladesh', *World Development* 66 (2015): 749–61. doi: 10.1016/j.worlddev.2014.09.022.

15 Julia Andrea Behrman, 'Do Targeted Stipend Programs Reduce Gender and Socioeconomic Inequalities in Schooling Attainment? Insights From Rural Bangladesh', *Demography* (2015): 1917–27.

launched combined health and family planning programmes with women as field work-ers.[16] It proved highly effective and birth rates dropped. In 2004 the UN predicted that by 2050 Bangladesh would have 255 million people, yet in 2012 its prediction has fallen to just 202 million because of rapidly falling birth rates. Two decades ago writers were predicting famine; now the prediction is for rice exports, because of the falling birth rate and rice production rising much faster than population.[17]

But there is another side to this rosy picture. Azizur Rahman Khan,[18] in a review of the Drèze and Sen book, notes that 'there has been a rather sharp increase in inequality in the distribution of income in Bangladesh over the last two decades, a period in which it has also much improved its social indicators that Drèze and Sen praise. The sharp increase in inequality has largely been due to public policy that encouraged concentra-tion of income and wealth at the top of the distribution.'

Both pictures of Bangladesh are true, and present contradictions that we will show have a real impact on the response to climate change. In rural areas and smaller cities, pro-gressive changes from improved rice varieties to cyclone protection systems have improved conditions and are becoming part of the response to climate change. But in the two big-gest cities, Dhaka and Chittagong, privatization and pursuit of wealth, by means legal and illegal, are making it much more difficult to respond to impacts of climate change.

What Is the Risk?

Bangladesh is often loosely cited as one of the countries most vulnerable to climate change, which seems intuitively reasonable as it is quite flat and susceptible to floods, sea level rise and cyclones. But if we look at estimates that are based on models and data, the picture is confusing. In 2011 David Wheeler of the Centre for Global Development saw Bangladesh as the fourth most threatened country – but it followed three African countries highly susceptible to drought and crop failure.[19] For Wheeler, the biggest risk to Bangladesh was crop loss, which, as we note in Chapter 2, is not seen by other experts as being the most serious problem. In contrast, the *World Risk Report 2015* puts Bangladesh sixth after Guatemala, the Philippines and three other island states. Its 'exposure' to risk puts it in tenth place and it is not even in the top 15 for 'vulnerability'. Its high risk is partly due to 'lack of coping capacity', which in turn is based on a perception of poor governance.[20]

16 Kenneth R. Weiss, 'How Bangladesh's Female Health Workers Boosted Family Planning' (London: *Guardian*, global development blog, 6 June 2014).

17 Mohammed Mainuddin and Mac Kirby, 'National Food Security in Bangladesh to 2050', *Food Security* 7 (2015): 633–46. doi: 10.1007/s12571-015-0465-6.

18 Azizur Rahman Khan, 'Living Standards, Inequality and Development: Some Issues with Reference to Comparisons between India and Bangladesh', *Journal of Human Development and Capabilities* 15 (2014): 429–36, doi: 10.1080/19452829.2014.966966.

19 David Wheeler, *Quantifying Vulnerability to Climate Change: Implications for Adaptation Assistance* (Washington, DC: Centre for Global Development, Working Paper 240).

20 *World Risk Report 2015* (Berlin: Bündnis Entwicklung Hilft and Bonn: United Nations University – EHS, 2015).

Many of the Bangladeshis we interviewed challenged the view that their country lacks coping capacity, and this book is largely about how Bangladesh is coping and adapting. Bangladesh has serious problems with corruption and privatization of public spaces, and the legal and illegal power of what are called 'influential people'. But we have also seen that in the event of a flood or a cyclone, personal interests are temporarily put aside and the country can pull together and respond. Similarly there is impressive development of agriculture, cyclone warning systems and other things that are essential to respond to climate change. Yet, we also see a division. Effective adaptation is taking place in rural areas and small cities. But the legal and illegal privatization of urban spaces, especially illegal buildings, the infilling of canals and lakes and the lack of public transport, is creating huge vulnerabilities. Dhaka and Chittagong probably lack the capacity to cope with a major flood similar to the one in 1998; the country's growing urban wealth is not building adaptive capacity in urban areas.

In this context it is interesting to look at a survey of the rich and powerful carried out by the World Economic Forum (WEF), which is most famous for its annual meeting in Davos, Switzerland. In 2014 it did a survey of almost nine hundred 'leading decision-makers', and asked them about risk. One question was 'For Which Global Risks Is Your Region Least Prepared?' and for the decision-makers from south Asia, it was the failure of urban planning, notably poorly planned cities and urban sprawl. The question was not specifically linked to climate change, but the problem is recognized, and for Bangladesh we see the biggest problems of climate change in the big cities.[21]

Climate change was seen as a risk in earlier WEF surveys, but it jumped to the top of the 2015 survey of 750 leading decision-makers. The risk with the most impact is that 'Governments and businesses fail to enforce or enact effective measures to mitigate climate change, protect populations and help businesses impacted by climate change to adapt.'[22] And they are right to be worried. The WEF survey was released just after the December 2015 climate change talks in Paris, known as COP 21. In Paris 'the international community not only acknowledged the seriousness of climate change, it also demonstrated sufficient unanimity to define it quantitatively' as an increase in global temperature since the start of industrialization 'well below 2°C' and preferably 1.5°C, pointed out Prof. Kevin Anderson, deputy director of the Tyndall Centre for Climate Change Research at the University of Manchester.[23] Unfortunately, he added, the decisions at Paris 'eliminates any serious chance' of keeping the temperature rise below 2°C.[24]

In this book we argue that Bangladesh is already adapting to climate change and has taken a lead in international negotiations to try to force countries to curb their carbon

21 World Economic Forum, *Global Risks 2015* (Geneva: World Economic Forum, 2015).

22 World Economic Forum, *Global Risks Report 2016* (Geneva: World Economic Forum, 2016), 5, 85.

23 Kevin Anderson, 'Talks in the City of Light Generate More Heat', *Nature* 528 (2015): 437.

24 Kevin Anderson, 'The Hidden Agenda: How Veiled Techno-Utopias Shore up the Paris Agreement', 2015. Accessed 6 September 2016, http://kevinanderson.info/blog/wp-content/uploads/2016/01/Paris-Summary-2015.pdf.

emissions. But as the 2015 Paris conference showed, this is not enough. Bangladesh will need to do much more to keep its head above water – both on the ground at home and in international forums.

In Chapter 2, we look at the projections for climate change in Bangladesh. Bangladesh already deals with an extremely complex environment, and climate change adds nothing new – but it makes existing problems substantially worse. There will be more damaging cyclones and floods, more rain and sea level rise. Bangladesh can adapt to an increased global temperature of 1.5°C, but higher temperatures pose serious problems later in this century.

In 2009 Bangladesh became the first developing country to frame a coordinated action plan, the Climate Change Strategy and Action Plan (BCCSAP). Chapter 3 shows how Bangladeshi scientists and activists have taken a leading role in promoting the climate change issue at home and internationally and how they have played an increasingly central role in international negotiations. After two decades, the industrialized world is accepting that climate change is real and that substantial damage has been caused.

Key to keeping Bangladeshi heads above water is responding to sea level rise. Chapter 4 describes the realization that this delta, with a billion tonnes of sediment pouring down from the Himalayas each year, can actually raise the land level to match sea level rise. But it has taken four decades for communities drawing on traditional knowledge to show the experts how it is done.

Perhaps the most dramatic success, almost totally unrecognized outside the country, has been the way independent Bangladesh has used cyclone shelters and early warning systems to cut the death rate from super cyclones to just 2 per cent of their former levels. But as Chapter 5 notes, climate change will require more shelters and stronger dykes.

River floods can be a blessing, bringing renewal to the land, but in the worst years can be devastating, and climate change will increase the number of bad floods. Debate about dealing with river floods and misunderstandings of outside agencies is the subject of Chapter 6.

It may be densely populated, but Bangladesh feeds itself. In part this is due to a dramatic fall in both the birth rate and therefore in population growth. The other reason is a total transformation of rice farming in just three decades. Chapter 7 describes what more will be needed to be done to deal with flood, salinity and other issues to keep rice production increasing in the face of climate change.

A widely propagated myth is that there are already climate change refugees in Bangladesh. Floods, erosion and cyclones have forced people to move for centuries, and they might be called 'environmental migrants', although they tend to move short distances. But there has been a huge migration to the cities, largely driven by poverty and a desire for better jobs. Poverty has been worsened by environmental factors, and climate change will further exacerbate such problems. Chapter 8 notes that even though there are no climate change migrants yet, there surely will be.

Bangladesh is still predominantly rural and the big environmental and climate change successes are there. In contrast, Dhaka, a megacity of 17 million, cannot even deal with ordinary rainfall and is in no position to cope with climate change. Half the population lives in slums, which the urban elites try to ignore. Chapters 9 and 10 report on the urban crisis, the limited actions that individuals and communities take to respond and the real threat of climate change.

Bangladesh took the lead in setting up a climate adaptation fund with its own money and tried to steer a path not determined by donors. Politics and mismanagement weakened the local hand, while the big donors took a hard line that outsiders know best. Chapter 11 is about the ongoing struggle.

In Chapter 12, we wrap up the book.

Temperature, sea level and rainfall are all increasing because of two centuries of industrialization in the rich countries. But rain, floods and heat are not new, and coping with this delta's complex and difficult ecology has made Bangladeshis experts in response and adaptation. This nation is not responsible for climate change but it will pay the price. But Bangladeshis will not accept the role of victims. Scientists, engineers, political leaders and community groups are already making the changes needed to cope with climate change. Much more will be needed and the countries and citizens of the rich world must take on their responsibilities to mitigate carbon emissions and finance climate change adaptation.

Chapter Two

HOW WILL CLIMATE CHANGE
HIT BANGLADESH?

Choices made at the Paris climate change conference in December 2015 and actions after that will determine if Bangladeshis can keep their heads above water, or must flee to high ground. But the real impact of the decisions will only be felt in 50 to 100 years because the fossil fuels we use today and in the near future create 'greenhouse gases', which persist in the atmosphere for decades, continuing to warm the planet and having an impact on future generations.

Emission limits proposed by Bangladesh and most developing countries would cause global warming to peak in the middle of this century, followed by a slow fall in temperatures. Bangladesh could cope with those changes. But the industrialized countries, and China and India, do not accept these limits, and the Paris conference accepted voluntary limits that will allow temperatures to rise well into the twenty-second century. In Bangladesh, this would cause flooding and migration, disrupted weather patterns and more severe storms, and possible food shortages. Protecting Bangladesh from climate change will require much more ambitious commitments from the industrialized – and industrializing – nations.

Perhaps the worst problem is that elected politicians are being asked to make hard choices that will only benefit their great-grandchildren but will impose cost on present-day taxpayers. In 2015, British prime minister David Cameron made sweeping cuts to programmes to reduce greenhouse gases, arguing that cutting immediate government spending was more important, while a leading US presidential candidate did not even accept the existence of human-created climate change. For decades to come, Bangladesh will keep its head above water only if it can be part of an international coalition maintaining enough pressure on global politicians to keep their initial very limited promises, and to make further commitments to cut emissions.

Predicting the future is hard, but the past decade has seen huge amounts of scientific research and major improvements in the models used to make projections of the impact of global warming. The Intergovernmental Panel on Climate Change (IPCC), established jointly by the United Nations Environment Programme (UNEP) and the World Meteorological Organization (WMO), has been collating the research and provides the most widely accepted and clearest scientific view on what is known – and not known – about climate change. It has three working groups involving hundreds of scientists, and its *Fifth Assessment Report* was published in sections in 2013 and 2014.[1] In the rest of this

---

1   Intergovernmental Panel on Climate Change (IPCC), *Fifth Assessment Report* (Geneva: IPCC and New York: Cambridge University Press, 4 volumes, 2013–14), http://ar5-syr.ipcc.ch/.

chapter, we use this report and other research to consider the impact on Bangladesh of four different levels of greenhouse gas emissions.

Global Climate Change

The earth's climate is hugely variable and also follows cycles ranging from annual (winter-summer) to decades, centuries or thousands of years long. These are linked to El Niño, sunspots, and variations in the way the earth spins around its axis and orbits around the sun. Other factors such as volcanoes can affect climate. The big difference now is that the change is mainly human-induced, which combines with the different cycles.

There is a broad scientific consensus that our world is getting warmer and this is mainly caused by gases emitted by the fossil-fuel-based industrialization and development processes. These gases collect in the atmosphere and act like the glass roof of a greenhouse (glasshouse) allowing solar radiation to pass but trapping heat below. Thus these are known as 'greenhouse gases'. Since the start of the industrial revolution, the average surface temperature of the earth has increased by 1°C. This is already having an effect on weather, agriculture and our lives, and is known as 'anthropogenic climate change'. The most important greenhouse gas is carbon dioxide (CO_2), 86 per cent of which comes from burning fossil fuels and industrial processes. Methane (CH_4), nitrous oxide (N_2O), sulphur hexafluoride (SF_6) and two groups of gases, hydrofluorocarbons (HFCs) and perfluorocarbons (PFCs), are the other important greenhouse gases.[2]

Climate change has varied effects. About 30 per cent of additional CO_2 is absorbed by the oceans, making them more acidic. A similar amount is absorbed by plants and increased CO_2 actually stimulates plant growth. The rest of the CO_2 remains in the air, increasing global warming, which melts the glaciers and ice sheets and adds water to the oceans. Meanwhile, as water gets warmer, it expands. Together these raise the sea level. Rising temperatures often restrict plant growth, frequently cancelling out any gains from the extra CO_2. Rising temperatures also make weather more erratic and make extreme events like droughts and cyclones more severe. If the temperature continues to rise unchecked, there will be massive coastal flooding, affecting many of the world's major cities, and there will be food shortages.

The problem was recognized in the late 1980s as it became clear that the continued increase in the use of fossil fuels such as coal and oil, and thus of greenhouse gas emissions, would have a catastrophic impact. The first major international meeting was the Earth Summit in Rio de Janeiro in 1992, where the United Nations Framework Convention on Climate Change (UNFCCC) was agreed. It sets no limits on greenhouse gas emissions for individual countries, contains no enforcement mechanisms and merely provides a framework for negotiating. The signatories to the Convention have met annually since 1995 as the Conferences of the Parties (COP). At COP 3 in Kyoto, Japan, in 1997, the Kyoto Protocol was adopted and 37 industrialized countries, including the European Union, but not the United States, committed themselves to limited but binding targets for greenhouse gas emissions.

---

2   IPCC, *Climate Change 2014: Synthesis Report* (Geneva: IPCC, 2014), 2–5.

Over the past three decades there has been massive research on climate change. Computer-based models combine the growing amount of weather data with equations that explain what we know about how climate and weather works at global and, to a lesser extent, local level. In the past decade a large number of these models have simulated aspects of past and future climate change, with increasing accuracy. This research is pulled together by the IPCC, which provides the most authoritative estimates and projections[3] of climate change. Its Fourth and Fifth Assessment Reports were published in 2007 and 2013/14. Some of the models contain more uncertainty than others, and the Fifth Assessment Report is very careful to estimate how confident it is are about projections and to give a range of estimates; we have simplified the estimates here and readers who want more details should refer to the IPCC reports. This chapter summarizes some of the IPCC conclusions, especially as applied to Bangladesh.

Projections of two numbers are key: global warming as measured by temperature rise and the resulting increase in sea level. Estimates of global warming are usually of the increase in temperature since the beginning of industrialization. These are the headline figures. IPCC estimated that between 1850 and 2012 temperatures had increased by 0.85°C,[1] but the most recent 'official' figure comes from the WMO and is that by 2015 temperature had increased by 1°C.[5] Thus the headline figure of a 2°C increase by 2100 is the same as an increase of 1°C during the rest of the twenty-first century. In contrast, sea level rise tends to be measured only from 2000, because earlier measurements are not very good. (However, based on more limited data, the IPCC estimates that sea level has risen by about 20 cm between 1900 and 2015.)

Climate Change and Bangladesh

Although Bangladeshis traditionally talk of six 'seasons', in practical terms there are only four (Table 2.1): the dry winter season from December to February, the pre-monsoon hot summer season from March to May, the rainy monsoon season from June to September and the post-monsoon autumn season of October and November.[6] More than 75 per cent of the rainfall in Bangladesh occurs during the monsoon season. Adverse weather

---

3   IPCC prefers to use the word 'projection' rather than 'prediction' to emphasize that these are based on models and contain uncertainty about assumption and the way processes evolve. For a discussion of this, see Dennis Bray and Hans von Storch, ' "Prediction" or "Projection"? The Nomenclature of Climate Science', *Science Communication* 30 (2009): 534–43.

4   The models used typically take 1750–1880 for the start of industrialization. IPCC estimates that by 2000, temperatures had already increased by 0.6°C and by 2012 had increased by 0.85°C. IPCC, *Climate Change 2014: Synthesis Report*, 2–5.

5   World Meteorological Organization, *WMO Statement on the Status of the Global Climate in 2015* (Geneva: World Meteorological Organization, 2016), 2.

6   Bangladeshis traditionally talk of the six seasons shown in the table, but it has been argued that in reality there are only four seasons. Moinul Islam and Koji Kotani, 'Six or Four Seasons? An Evidence for Seasonal Change in Bangladesh', Kochi University Technical Report, 2014, doi: 10.13140/2.1.3011.5845.

Table 2.1 Bangladesh weather and agricultural seasons

	December	January	February	March	April	May	June	July	August	September	October	November
Climate Season	Winter			Pre-Monsoon			Monsoon				Post-Monsoon	
Bengali Season	Sitkal Winter			Bosontokal Spring		Grissokal Summer		Borsakal Rainy season		Soratkal Pre-Autumn		Hemontokal Late Autumn
Weather	Cool, dry			Hot; variable rainfall			Monsoon (75% of rains)					
Shocks					Drought, cyclone			Flood			Cyclone	
Rice	Boro – irrigated; transplanted – now main rice crop				Aus – rainfed; transplanted lowland and direct seeded upland			Aman – flood and rainfed; transplanted & direct seeded				
							Deepwater – tall and floating for deep flooding; direct seeded – reducing in importance					
Wheat		Wheat										
Crop Seasons	Rabi			Kharif-1				Kharif-2				Rabi

events are seasonal. Floods occur during the monsoon season, cyclones during the pre- and post-monsoon seasons and drought during the pre-monsoon.[7] (See Table 2.1.[8] Rice seasons shown in the table are discussed in later chapters.)

Bangladesh may be small but it is so varied that it is almost impossible to make generalizations. The Asian Development Bank notes that although the average annual precipitation is 2,320 mm, this varies from 1,110 mm in the west to 5,690 mm in the northeast. Mean annual temperature is about 25°C, with extremes as low as 4°C and as high as 43°C.[9] Floods are caused by three major rivers and local rainfall, all of which are different each year. Moreover the country is an active delta that is constantly changing.

Normal flood and cyclone patterns in Bangladesh are too irregular to provide any evidence of climate change effects yet. Sea level rise is less precisely measured and in recent years appears to be somewhere between 3 and 10 mm per year.[10] Estimating this is difficult because most of Bangladesh is an active delta and coastal areas receive huge amounts of sediment, which not only raises land levels in some places (discussed more in Chapter 4), but also compacts over years in some areas leading to subsidence of perhaps 2–3 mm per year.[11] In some places islands vanish while in others new islands appear. The combined impact of subsidence, new sediment and rising sea levels has different effects at different points along the coast.

Bangladesh only has good quality weather records going back six decades. Also, there are problems that some weather stations have been moved while others that were once in rural areas have caught up by urbanization and urban areas have higher temperatures. Thus the data is subject to varied interpretations, and researchers reach different conclusions. Shamsuddin Shahid and Hugh Brammer both looked at the period 1958–2007. Shahid concluded that mean[12] temperature rose 0.01°C per year, or 0.5°C over a half century – perhaps a bit more than the global average. The temperature rise is largely in the warmer summer periods, and the winter is slightly colder, especially in the coolest

7 Climate Change Cell, Department of Environment, Government of Bangladesh, *Climate Change and Bangladesh* (Dhaka: Climate Change Cell, 2007), 5.

8 The table is partly drawn from Hugh Brammer, *Can Bangladesh be Protected from Floods?* (Dhaka: University Press, 2004), 41.

9 Mahfuz Ahmed and Suphachol Suphachalasai, *Assessing the Costs of Climate Change and Adaptation in South Asia* (Mandaluyong City, Philippines: Asian Development Bank, 2014), 24.

10 Abu Bakar Siddique, 'Sea Level Rising by 6–20 mm each Year', *Dhaka Tribune*, 30 October 2015, citing an unpublished report 'Assessment of Sea Level Rise and Vulnerability in the Coastal Zone of Bangladesh through Trend Analysis' (Dhaka: Centre for Environmental and Geographical Information Services). Unofficially, a CEGIS official said their estimate was 5–7 mm per year sea level rise in central coastal areas.

11 Hugh Brammer, *Climate Change, Sea Level Rise and Development in Bangladesh* (Dhaka: The University Press, 2014), 155.

12 There are two kinds of averages. 'Mean' is the most commonly used, where all the numbers are added up and then divided by the number of numbers. The 'median' is the middle value in the list of numbers listed in numerical order. So half of the values are above the median and half below. These are normally different. For example, in the mean income very large incomes have a great weight, so the mean income is always higher than the median income.

areas of the northwest.[13] Hugh Brammer also finds maximum temperatures in the cooler winter becoming lower and in the hotter monsoon and post-monsoon becoming higher. He reports that the cool winter period is becoming shorter and starting later.[14]

Shahid concludes that mean rainfall in Bangladesh increased by 5.53 mm per year, or 276 mm over the half century – an increase of 10 per cent. Rainfall increased primarily in the pre-monsoon season and in the northwest.[15] Brammer finds a general increase in rainfall except in the coastal areas, but there is no evidence so far of increased variability in rainfall.[16] Some researchers find indications of more extreme rainfall days.[17] However Shahid argues that 'the data and analysis in the present study are insufficient to draw conclusions about the impact of global climate change on Bangladesh' and Brammer stresses that the huge variability makes it impossible to link changes to global warming.[18]

Three Targets and Four Scenarios

Temperature increase is key, and there are three targets that have been put forward to limit global warming by 2100. We use IPCC models to create scenarios and projections of the impact based on 2100 global temperature projections. (The scenarios are explained in the footnotes.)

- *1.5°C – Bangladesh Proposal*: The UN's Least Developed Countries Group proposed at Copenhagen (COP 15) that temperature increase be limited to 1.5°C above pre-industrial levels (which is 0.5°C for the rest of this century). This was not accepted at Copenhagen but it is still the target of 106 of the 195 countries that went to Paris (COP 21) and of the 43-nation Climate Vulnerable Forum, whose spokesperson is a Bangladeshi, Saleemul Huq. We call this the Bangladesh Proposal scenario. IPCC model projections suggest that if global warming was kept to this level, then global temperature and CO_2 would peak in 2050 and then begin to decline slowly.[19]
- *2.0°C – Cancun*: At Cancun, Mexico (COP 16) in 2010, it was agreed to keep the increase in global average temperature below 2°C more than pre-industrial levels, and

---

13 Shamsuddin Shahid, 'Recent Trends in the Climate of Bangladesh', *Climate Research* 42 (2010): 185–93. Confirmed by Md. Abu Zafer Siddik and Mursheda Rahman, 'Trend Analysis of Maximum, Minimum and Average Temperatures in Bangladesh: 1961–2008', *Theoretical and Applied Climatology* 116 (2014): 721.

14 Brammer, *Climate Change*, 67, 89.

15 Shahid, 'Recent Trends', 185–93.

16 Brammer, *Climate Change*, 114–15, 127–28.

17 Faisal Ahammed, Guna Alankarage Hewa and John R. Argue, 'Variability of Annual Daily Maximum Rainfall of Dhaka, Bangladesh', *Atmospheric Research* 137 (2014): 176–82.

18 Shahid, 'Recent Trends', 191 and Brammer, *Climate Change*, xix–xx.

19 IPCC, *Climate Change 2013, The Physical Science Basis, Contribution of Working Group I to the Fifth Assessment* (Cambridge, UK and New York: Cambridge University Press, 2013). Estimates of temperature rise and CO_2 come from figure 12.40 (p. 1100) and figure 12.42 (p. 1103). IPCC Fifth Assessment Working Group I scenario RCP2.6 is the one with the lowest temperature rise and its projection is 1.6°C in 2100, so we use this for the Bangladesh proposal scenario. Note that in figure 12.40 RCP2.6 is labelled as RCP3-PD.

accepted that discussions should continue on reducing this to 1.5°C. At Paris (COP 21) this was made only slightly stronger: 'Holding the increase in the global temperature to well below 2°C above pre-industrial levels and to pursue efforts to reduce the temperature increase to 1.5°C above pre-industrial levels.' We call the 2°C increase by 2100 the Cancun scenario.[20] The projections of the models are that temperature and CO_2 would peak in 2100 and then remain constant.

- *2.7°C – Paris COP 21*: At Lima, Peru (COP 20) in 2014 it was agreed that countries would announce individual voluntary limits to greenhouse gasses before the Paris (COP 21) meeting in 2015. Before the COP 21, 147 countries representing 86 per cent of global greenhouse gas emissions in 2010 had submitted plans.[21] Christiana Figueres, executive secretary of the UNFCCC, said on 30 October 2015 that these plans 'have the capability of limiting the forecast temperature rise to around 2.7°C by 2100, by no means enough but a lot lower than the estimated four, five, or more degrees of warming projected by many' prior to these pledges.[22] COP 21 president, Laurent Fabius, the French minister of foreign affairs, proposed a paragraph in the final statement 'emphasizing with serious concern the urgent need to address the significant gap' between these pledges and the commitment to 'well below 2°C', but key industrialized countries would not allow this in the final agreement. Further it was agreed that there would be no enforcement of these pledges and that no new, lower pledges would be required until 2020. We call this the Paris COP 21 scenario.[23] The model projections are that CO_2 levels will increase until 2150, and that temperature would increase to 3.1°C by 2200 and continue to rise. Based on the Paris COP 21, this is now the most likely scenario.

- *Business as Usual*: The Paris pledges are voluntary and have no legal standing, so it seems likely that some countries will not meet their own targets, which would mean temperature rises of above 2.7°C. If there is no mitigation of greenhouse gas emissions, the

---

20 The IPCC does not have a scenario for 2°C, so we take the average of the 1.6°C scenario (RCP2.6) and the next IPCC scenario which projects 2.5°C at the end of the century (RCP4.5). IPCC's 2007 Fourth Assessment used different models, and model B1 gives temperature results similar to RCP4.5. Comparisons between fourth assessment and fifth assessment models are given in Figure 1–4, page 179, of IPCC, *Climate Change 2014: Impacts, Adaptation, and Vulnerability. Part A: Global and Sectoral Aspects. Contribution of Working Group II to the Fifth Assessment Report of the Intergovernmental Panel on Climate Change* (Cambridge, UK and New York: Cambridge University Press, 2014). Model B1 was used by the Asian Development Bank as their 'Copenhagen-Cancun scenario.' Ahmed and Suphachalasai, *Assessing the Costs*.

21 These plans are known as Intended Nationally Determined Contributions (INDCs).

22 'Global Response to Climate Change Keeps Door Open to 2 Degree C Temperature Limit', UNFCCC secretariat press release, Berlin, 30 October 2015. Accessed 8 September 2016, http://newsroom.unfccc.int/unfccc-newsroom/indc-synthesis-report-press-release/.

23 The IPCC does not have a scenario for 2.7°C, so we take a weighted average of the IPCC 2.5°C scenario (RCP4.5) and the next scenario, which projects 3.1°C in 2100 (RCP6.0). The IPCC's fourth assessment models A1B and B2 are similar to fifth assessment RCP6.0.

IPCC models project a 4.9°C rise by 2100 and 9°C by 2200, with CO_2 levels rising until 2250.[24] We call this the Business as Usual scenario.

The IPCC also estimates sea level rise. Table 2.2 summarizes the global projections.[25] Current sea level rise is caused by thermal expansion – warmer water takes up more space so the sea expands and pushes up the sea level. But by the end of this century, one-third of sea level rise will be due to melting glaciers. By the end of the next century, water from the melting of the Greenland ice sheet will play a major role.[26]

The projections of all models are that increased temperature triggers increased rainfall, and that the sea will become more acidic as it absorbs more CO_2.

Small Differences Have a Big Impact

Initially the forecasts in Table 2.2 appear relatively close, even at the end of this century, and 1.5°C does not seem all that different from 2°C or 2.7°C. But the IPCC fifth assessment finds that there are big differences even between the lower warming scenarios. Two aspects are important for Bangladesh. First, above a temperature rise of 1.5°C (the Bangladesh Proposal), there is a 'high' risk of extreme climate events such as heat waves and extreme precipitation – which will hit Bangladesh in particular. Second is what happens in the longer term, that is, into the twenty-second century; greenhouse gas builds up in the atmosphere and continues to cause problems for decades, dissipating only slowly. The projections of the Bangladesh Proposal scenario are that warming stops in the mid-twenty-first century, which means climate change damage can be managed in Bangladesh. But the other scenarios show damage worsening into the next century. The two middle range scenarios point to serious and increasing damages: the scenario with 2°C in 2100 finds that warming slows down only after 2100, and for 2.7°C finds warming does not slow down until it exceeds 3.1°C in 2200. In the fourth scenario, warming continues through the twenty-second century and reaches 9°C by 2200, which would be catastrophic.

As we will show later in this book, these differences matter. Bangladesh can cope with a 1.5°C temperature rise. But even the 2.7° increase accepted at the Paris (COP 21) meeting will have major sea level rise, flooding and increased food costs. So Bangladesh is right to fight for a much lower temperature increase.

Looking More Closely at Bangladesh

As well as doing global estimates, the IPCC does much more detailed regional projections. For Bangladesh, the IPCC's fifth assessment projects as follows:

---

24 IPCC scenario RCP8.5, which is halfway between the 4th assessment A2 and A1F1.
25 For given 2100 temperatures, the fifth assessment projections are for somewhat higher sea levels than the fourth assessment.
26 IPCC, *Climate Change 2013, The Physical Science Basis, Contribution of Working Group I*, tables TS.1 and 13.8 and figures 12.40 and 12.42, pp. 90, 1100, 1103 and 1191.

Table 2.2 IPCC projections of global increases in surface air temperature, compared to pre-industrial levels, and sea level compared to 1986–2005

Scenario	Temperature rise (°C) compared to 1850			Sea level rise (cm) compared to 1986–2005		
	2055	2090	2200	2055	2090	2200
Bangladesh proposal	1.6	1.6	1.4	24	40	53
Cancun	1.8	2.0	2.0	26	45	62
Paris COP21 pledges	2	2.7	3.1	26	48	88
Business as usual	2.6	4.9	9.0	30	63	130

Note: Our estimates from various IPCC charts and tables.
Sources: IPCC, *Climate Change 2013, The Physical Science Basis, Contribution of Working Group I to the Fifth Assessment Report of the IPCC* (Cambridge, UK and New York: Cambridge University Press, 2014), tables TS.1 and 13.8 and figures 12.40 and 12.42, pp. 90, 1100, 1103 and 1191.

- Temperature increase in Bangladesh will be slightly less than the global average.[27]
- Sea level rise in the Bay of Bengal will be close to the global averages.[28]
- Tropical cyclone frequency is likely to decrease or remain the same, but intensity (wind speed and rainfall) is likely to increase. The number of the most intense cyclones, which cause the worst damage, is likely to increase. There are also suggestions of higher storm surges and sea level extremes.[29]
- Rain and flood: There will be a greater than average increase in precipitation, which will slightly reduce the need for irrigation. The monsoon season could become longer. Up to 2100, rivers will experience increased flows of water due to both more rainfall and more melting of snow and glaciers in the Himalayas into the Ganges and Brahmaputra rivers. More severe flooding is expected, worsening as temperatures rise. No change is expected to the recharging of groundwater.[30]
- Food: In the Bay of Bengal, fish catch potential will increase substantially in some parts and decrease in others, forcing changes in fishing practices, but could overall increase yields.[31] Rice production will increase because rice yields increase with higher CO_2 in the atmosphere, while some increases in temperature cause no harm and with adaptation could actually increase yield. Wheat yields will fall substantially with any increase

---

27 IPCC, *Climate Change 2014: Impacts, Adaptation, and Vulnerability*, Working Group II, 10, 139.
28 IPCC, Climate Change 2013, The Physical Science Basis, Working Group I, 1198.
29 IPCC, Climate Change 2013, The Physical Science Basis, Working Group I, 1200–1201, 1249–50 and IPCC, Climate Change 2014, Impacts, Adaptation and Vulnerability, Working Group II, 147.
30 IPCC, Climate Change 2013, The Physical Science Basis, Working Group 1, 1273 and IPCC, Climate Change 2014, Impacts, Adaptation and Vulnerability, Working Group II, 66, 159, 191, 242, 248, 250.
31 IPCC, Climate Change 2014, Impacts, Adaptation and Vulnerability, Working Group II, 69, 458.

in temperature[32] although there is research underway into more heat tolerant varieties of wheat.

- Coastal mangroves are considered at high risk from sea level rise, increased salinity and storms. Mangroves provide an important coastal protection and the IPCC stresses the need for coastal reforestation.

The Asian Development Bank (ADB) issued a report on South Asia in 2014. Key points include:

- Cost: Climate change losses will cost Bangladesh 1 per cent of GDP by 2050 and 2 per cent of GDP by 2100 under the Cancun scenario, but 2 per cent of GDP by 2050 and 9 per cent of GDP by 2100 under the Business as Usual scenario.[33]
- Temperature increases in Bangladesh will be less than in neighbouring parts of India.[34]
- Total rainfall will increase about 10 per cent later in the century; not much change in drought patterns is expected. Increased rainfall will significantly increase Bangladesh's total water supply – in sharp contrast to neighbouring India, which will suffer major deficits and where demand will exceed supply by 2030.[35]
- Food: Increased production of rice and decreased wheat production is confirmed. For higher temperature rises the productivity gains are for the dominant boro (dry season irrigated) rice in the north, and there will be productivity falls in the west and south.[36]

Further projections on rainfall and river flooding were published in 2015 using models of the UK Met Office Hadley Centre for Climate Science. Rainfall increase is confirmed but there is a warning of more 'extreme events' and an 'increased risk of flash flooding'. Projections suggest a general decrease in the frequency of days with light or moderate rainfall (up to 12.3 mm per day) but 'a large increase in very heavy events' (more than 23.8 mm per day).[37] By the end of the twenty-first century the dry season is expected to be two to three weeks longer than now, concentrating more rainfall in a shorter period.[38] The other projection is that river flooding will worsen by 2050 due to increased monsoon season flows in the three major rivers. 'The extreme high flow exceeded 5 per cent of

---

32 IPCC, Climate Change 2014, Impacts, Adaptation and Vulnerability, Working Group II, 497–99.

33 Ahmed and Suphachalasai, *Assessing the Costs*, 74–80. ADB projections are based on the IPCC Fourth Assessment Report.

34 Ibid., 20.

35 Ibid., 23 (table 2), 69–72.

36 Ibid., 51–52, 83.

37 J. Caesar et al., 'Temperature and Precipitation Projections over Bangladesh and the Upstream Ganges, Brahmaputra and Meghna Systems', *Environmental Science Processes and Impacts* 17 (2015): 1047, 1054. The Met Office Hadley Centre used several models. One gave a projection for a slight fall in rainfall in 2014–60, but then a large increase by 2080–99. The others generated projections of a large increase in both periods.

38 D. Clarke et al., 'Projections of On-Far Salinity in Coastal Bangladesh', *Environmental Science Processes and Impacts* 17 (2015): 1127, 1131.

the time is projected to increase to 15 per cent by mid-century.' But later in the century dry season flows are likely to decrease. However, one study warns that this could change if new dams were built and large amounts of water were extracted upstream from the Brahmaputra and Ganges; this could reduce flows during the entire year and cause a significant reduction in the lowest flows.[39]

The consensus that global warming will bring more rain and more floods partly depends on what will happen upstream on the Ganges and Brahmaputra rivers. Monsoon rain in India is important, but glaciers also feed the rivers. The Brahmaputra receives 40 per cent of its water from mountains above 2,000 m in India and China – 27 per cent from glacier melting and snow and 13 per cent from rain. The Ganges receives only 11 per cent of its water from mountains in India and Nepal, nearly all from glaciers and snow.[40] The Himalayan glaciers are already melting, which increases the flow of water, but at some point the glaciers will be sufficiently smaller so that the their contribution to the flow will decrease, probably before 2050.[11] After that, glacier shrinkage will depend on carbon emissions; by the end of the century more than one-third of the glaciers in the headwaters of the Ganges will be gone under the Paris COP 21 Pledges scenario, and at least half will have melted away under the Business as Usual scenario. However, climate change is also expected to substantially increase precipitation in the high Himalayas. Water from increased snow melt and rainfall will increase the river flows and that will continue even after the glacier melt starts to decrease.[12] Increased precipitation due to global warming will be during the monsoon and more of the precipitation will be rain rather than snow due to rising temperatures,[13] which will increase monsoon flooding. Dry season flows will have two contradictory influences, and will be decreased by the shift in rainfall, but increased by the rise of the water table in the catchment areas continuing to release water during the dry season.

The Sundarbans, the largest single block of tidal mangroves in the world, are one of Bangladesh's most sensitive ecosystems,[11] and the models have difficulties taking in a large range of effects.

It is estimated that the 45 cm additional sea level rise expected even under moderate global warming (see Table 2.2) could inundate 75 per cent of the Sundarbans,[45] and reports say sea level rise has already submerged some low-lying islands and sandbars.

---

39 P. G. Whitehead et al., 'Impacts of Climate Change and Socio-Economic Scenarios on the Flow and Water Quality of the Ganges, Brahmaputra and Meghna (GBM) River Systems', *Environmental Science Processes and Impacts* 17 (2015): 1057, 1065–67.

40 Walter W. Immerzeel, Ludovicus P. H. van Beek and Marc F. P. Bierkens, 'Climate Change Will Affect the Asian Water Towers', *Science* 328 (2010): 1382–85.

41 W. W. Immerzeel, F. Pellicciotti and M. F. P. Bierkens, 'Rising River Flows Throughout the Twenty-First Century in Two Himalayan Glacierized Watersheds', *Nature Geoscience*, 6 (2013): 744, doi: 10.1038/NGEO1896.

42 Immerzeel, Pellicciotti and Bierkens, 'Rising River Flows' predicts increased runoff into the Ganges 'at least until 2100'.

43 Ahmed and Suphachalasai, *Assessing the Costs*, 4, 36.

44 More than two-thirds of the Sundarbans are in Bangladesh and the rest in India.

45 IPCC, *Climate Change 2001, Impacts, Adaptation and Vulnerability*, Contribution of Working Group II to the Third Assessment (Cambridge, UK and New York: Cambridge University Press,

But as we note in Chapter 3, the sediment that comes down the rivers is creating new sandbars and in many places keeping up with sea level rise (see Box 4.1). Increasing salinity is already leading to a natural shift to smaller but more salt-resistant mangrove species[16] but the ADB study warns that increased freshwater reaching the sea in monsoon floods could have 'significant consequences for the composition of the Sundarbans mangroves'.[17] The Sundarbans provide an important natural buffer, slowing down cyclones before they make landfall, and there is no agreement as to how the various influences of climate change will combine there. But the fear is that not only will cyclones be more severe, but they will also be less attenuated by smaller salt-tolerant trees that could make the Sundarbans a less effective buffer.

Conclusion: More of the Same … but How Much More?

Climate change will not create new problems for Bangladesh, but it will exacerbate many of the existing problems. A country known for its variable and extreme weather is likely to have more unpredictable and more extreme weather. Thus the number of cyclones will not increase, but some of them will be much stronger and more damaging. Rainfall will increase, but largely in the form of very heavy and possibly damaging downpours. River flooding will increase. As we note in subsequent chapters, Bangladesh has experience coping with these severe events but climate change will require more protection, more damage repair and improved systems to help affected people. The ADB predicts that costs will be 2 to 9 per cent of GDP – a vast amount of money.

Rice production will continue to increase, although further adaptations will be required. But wheat may become unviable as a domestic crop, and most wheat will need to be imported.

Sea level rise will not be as apocalyptic as earlier commentators in the climate change debate claimed; Bangladesh will not disappear. But the southwest and particularly the 10 per cent of Bangladesh that is less than 2 m above sea level will be particularly hard hit by rising sea levels and associated salinization. Even this will require major changes to the dyke and polder systems (discussed in Chapter 4) and to local farming systems. The projections of all models are that if global warming continues to levels that cause melting of the Greenland and Antarctic ice sheets, then in coming centuries sea levels will rise by several metres.[48]

All of the various models make similar projections for the next 40 years, and Bangladesh will be able to cope, albeit by incurring massive additional costs. But failing to curb greenhouse gases will have dramatic impacts in the second half of this century

2001), 19.3.3.5. Mangrove Ecosystems. http://www.grida.no/climate/ipcc_tar/wg2/670.htm#19335.

46 Anirban Mukhopadhay et al., 'Changes in Mangrove Species Assemblages and Future Prediction of the Bangladesh Sundarbans Using Markov Chain Model and Cellular Automata', *Environmental Science Processes and Impacts* 17 (2015): 1111, 1116.

47 Ahmed and Suphachalasai, *Assessing the Costs*, 38.

48 IPCC, Climate Change 2013, The Physical Science Basis, Working Group I, 1191.

and in the twenty-second century. The temperature differences may seem small, but because of the persistence in the atmosphere of greenhouse gases the differences in effect are long term and significant. A global temperature rise limited to 1.5°C since the start of industrialization leads to a peak of temperature and greenhouse gases approximately forty years from now and then a decrease. This is the target being pushed by Bangladesh and other least developed countries, because they can cope with 1.5°C. But the industrialized and newly industrializing countries are pushing for increases of 2°C or even 2.7°C by the end of this century, which the models show leads to more than 3°C in the next century. That will be catastrophic for Bangladesh. Decisions made now will only really have an impact on our children, grandchildren and great-grandchildren. Are short-term priorities so important that we will let them drown?

Chapter Three

TAKING THE LEAD IN NEGOTIATIONS – AND MOVING FORWARD

Because climate change accentuates existing environmental problems in Bangladesh, its scientists could see clearly what was coming, and they have taken a leading role in the annual climate change negotiations, the COP talks, and in putting pressure on the industrialized countries. Their stress has been on two issues, pressing for the least possible temperature rise and putting on to the agenda the industrialized countries' responsibility for the damage they have already caused. As a new country but with an ancient culture that honours learning, Bangladesh put together knowledgeable, dedicated and hard-working negotiating teams. They played a key role in the interminable backroom negotiations, forcing the industrialized nations to take seriously the expertise from developing countries and pushing into agreements the key phrases that mean industrialized countries must recognize their responsibility for global warming.

Annual COP meetings are presented by the media in binary terms, as a failure or as a breakthrough. And to some extent they are. Meetings run all night and into extra days as world leaders struggle to reach compromises to curb climate change while protecting their national interests. But in this very technical arena, it is expert teams who lay the groundwork for the final compromises. And the agenda is set in the permanent, ongoing technical meetings where, very quietly, opinion is changed – ideas that seemed impossible a few years earlier come to be seen as reasonable and normal. This reflects better science and modelling, making clear the dangers of unabated climate change, as well as improved understanding of how emissions can be cut. The Bangladeshi expert negotiators have taken a major role and added significantly to this incremental progress.

Making an International Mark

In 1986 Saleemul Huq and others set up the Bangladesh Centre for Advanced Studies, in part to research the environment, and were joined by Professor Atiq Rahman. Major floods in 1987 and 1988 hit Bangladesh just as pioneering research on sea level rise was being published by the Woods Hole Oceanographic Institution, Massachusetts, US[1]. Huq and Rahman realized just how serious the impact on Bangladesh might be – 'all of

---

1   F. J. Gable and D. G. Aubrey, 'Potential Coastal Impacts of Contemporary Changing Climate on South Asian Seas States', *Environmental Management* 14 (1990): 33–46.

Bangladesh might be under water, and we had better worry', Rahman said. A first book discussing this was published in 1990.[2]

Four other Bangladeshi intellectuals became key players, M. Asaduzzaman, from the Bangladesh Institute of Development Studies (BIDS), where he became research director; Z. Karim, who later became head of the Bangladesh Rice Research Institute and chair of the Bangladesh Agricultural Research Council; Qazi Kholiquzzaman Ahmad, also from BIDS but later chair of Dhaka School of Economics; and Professor Ainun Nishat, a pioneering expert of water resource management.

Nishat admits he was a reluctant convert. Sitting in his office as vice chancellor of BRAC University, he told us that in the 1980s he was a professor at the Bangladesh University of Engineering and Technology (BUET) and 'then, I did not believe in climate change'. Asked to include climate change in a report he was writing for the United Nations Economic and Social Commission for Asia (ESCAP), he admits he played it down. The 2001 Intergovernmental Panel on Climate Change (IPCC)[3] *Third Assessment Report* convinced him. Nishat was not alone in his initial scepticism. In the early 1990s BUET refused to allocate time for climate change research because it was 'science fiction'.

For the 1992 Rio Earth Summit, Bangladesh sent a very large delegation, including three ministers. But Nishat notes that 'climate change was still seen as something pushed by the developed world. And civil servants just followed the G77 and China'.[4] Subsequent Bangladesh governments and the political establishment were not interested in climate change.

Despite the lack of local interest, Bangladeshi climate change scientists became central to the climate change debate and they published key books in the late 1990s[5] and received more recognition internationally. Atiq Rahman told us, 'We started with a level playing field – climate change was new to everyone. Our advantage was that we had knowledge from experience, knowing what would happen.' Although much of the research was being done in the South, the analysis and publications were in the North. 'So in the IPCC we fought hard to bring in voices other than just from the United States and Europe.'

---

2 Atiq Rahman, Saleemul Huq and Gordon Conway, eds, *Environmental Aspects of Surface Water Systems of Bangladesh* (Dhaka: University Press, 1990).
3 The IPCC was established in 1988 by two United Nations organizations, the World Meteorological Organization (WMO) and the United Nations Environment Programme (UNEP). IPCC produced its first assessment report in 1990. The United Nations Framework Convention on Climate Change (UNFCCC) is the international environmental treaty negotiated at the Earth Summit in Rio de Janeiro in June 1992, and the IPCC became the main scientific body supporting the UNFCCC. It assessment reports are seen as the best scientific consensus on climate change and it is now the main international authority on climate change.
4 The G77 is the Group of 77 at the United Nations and is a loose coalition of developing nations, originally with 77 members, which later grew to 134.
5 R. A. Warrick and Q. K. Ahmad, *The Implications of Climate and Sea-Level Change for Bangladesh* (Dordrecht, Netherlands: Kluwer, 1996) and S. Huq, Z. Karim, M. Asaduzzaman and F. Mahtab, eds, *Vulnerability and Adaptation to Climate Change in Bangladesh* (Dordrecht, Netherlands: Kluwer, 1998)

The Bangladeshis were a heavyweight team with British PhDs and serious research profiles, and they made their presence felt on the academic and campaigning side in the early 2000s. In 2001 Saleemul Huq set up the International Institute for Environment and Development (IIED) climate change programme based in London, and is now also director of the International Centre for Climate Change and Development (ICCCAD) in Dhaka. At the COP 21 talks in Paris in 2015, he was the spokesperson for the influential 43-nation Climate Vulnerable Forum – an informal negotiating group of the countries that will be most damaged by climate change.

Three of these academic pioneers worked on IPCC reports. Huq and Ahmad were lead authors of chapters in the 2001 *Third Assessment Report* and Huq, Ahmad and Rahman were lead authors of chapters of the *Fourth Assessment Report* in 2007. In 2008 Rahman was named by the United Nations Environment Programmes (UNEP) as one of its 'Champions of the Earth', because as executive director of BCAS he has 'transformed the NGO into a leading think-tank in South Asia on sustainable development issues'.[6]

Because the older generation of climate experts are linked to independent research centres, they have trained a next generation who have moved into strategic positions in technical teams of developing countries, carrying out preparations and then providing backup for the talks.

Changing Governments Mean Changing Approaches

A major change came with the caretaker government of 2006–8.[7] C. S. Karim, a nuclear scientist, former international nuclear inspector for the International Atomic Energy Agency and chair of the Bangladesh Atomic Energy Commission, was nominated Agriculture, Fisheries, Livestock, Environment and Forests minister, which included climate change. One of his tasks was to draft the national paper for the Bali, Indonesia COP 13 talks in December 2007. He talked to us in his flat in Dhaka, serving us fruit he had grown on his roof[8] – Dhaka has few parks, but there are a surprising number of small trees on balconies and roofs. He smiled as he said, 'I committed a crime. I chose my own people and did not bother if they were in the ministry or not. If people are committed, it will work; if only self interest, it does not. I needed to choose people who could talk sense and already knew the subject.' In the end it was a mixed team with Nishat, Asaduzzaman and other independent scientists as well as ministry people. C. S. Karim then moved on to other climate issues, including making Bangladesh the first developing country to frame a coordinated action plan, the Climate Change Strategy and Action

6   'Climate Change Links 2008 Champions of the Earth Award Winners', UNEP press release, 28 January 2008. Accessed 10 September 2016, http://www.unep.org/documents.multilingual/default.asp?documentid=525&articleid=5738&l=en.
7   Under the then constitution, a caretaker government managed the country during the 90-day election period. It assumed power in October 2006. After arguments between the two main parties and boycott threats, the army intervened and the caretaker government was extended. Elections were held on 29 December 2008, and were won by the Awami League.
8   C. S. Karim died on 20 November 2015.

Plan (BCCSAP) – with Nishat, Asaduzzaman and Rahman on the team. In September 2008 Bangladesh and the UK jointly organized the London Climate Change Conference, but C. S. Karim insisted that it was the Bangladeshis who produced the national paper, without the involvement of the UK Department for International Development (DfID)

The 29 December 2008 election was won by the Awami League and Sheikh Hasina became the new prime minister,. Her new Environment minister, Hasan Mahmud, updated and published the BCCSAP and kept and expanded the same COP negotiating team. The COP talks are long and complex, the parallel sessions often going late into the night with arcane arguments about wording. European countries and the United States can throw large, experienced and well-briefed teams at these talks, and they can argue all night to prevent the inclusion of words that might admit their responsibility for greenhouse gases. Most developing countries send small teams of lawyers and government functionaries who are quickly overwhelmed by the process, and sometimes spend their time shopping. 'Bangladesh sends a large delegation, with a mix of academics and hard negotiators,' noted Asaduzzaman. The teams have even included a journalist, Quamrul Chowdhury, the founder of the environmental journalists' forum, to ensure that documents are readable and redrafting is coherent. One member of the team commented on the 'good collaboration among the government and non-government actors who collectively worked as members of the Bangladesh delegation'. And the negotiating teams work hard, meeting for breakfast and in the late evening to discuss tactics – where to fight and where to concede. Negotiations often run over time. At the Durban COP 17 in 2011, negotiations continued for two extra days. Many developing country teams went home on schedule, but the entire Bangladesh team stayed – and won some last-minute concessions. 'We were organized and effective,' commented Q. K. Ahmad, who became head of the negotiating team. Environment Minister Mahmud was criticized for also including in the COP delegations a group of political nominees who were really just travelling for a holiday or shopping, but they did not interfere with the negotiations and their presence may have been necessary to buy off influential people who might otherwise have tried to cut back on the professional team.

Bangladesh soon became a major player in international negotiations. At COP 16 in Cancun, Mexico, in 2010, Bangladesh Environment Minister Hasan Mahmud was co-chair (with Australia) and played a key role in winning developed country agreement for a Green Climate Fund. Bangladesh has also used its influence behind the scenes. A Ministry of Environment official, who asked not to be named, complained that India and China saw the right to develop as the right to catch up to the already developed countries and to pollute as much as they felt necessary. 'We were trying to get agreement in Durban and we wanted India on board. I told them "This is in your interest. If the sea level rises and Bangladesh is flooded, where do you think our people will go."'

Although the Bangladesh delegations have been an impressive mix of government and civil society, the gender mix has not been as good. The delegation to the COP 19 in Warsaw in 2013 had 70 people, of whom only 10 were women. Men dominate the negotiations, notes Sharaban T. Zaman, a law lecturer at Daffodil International University and NGO delegate at the COP talks. 'In UNFCCC negotiation it is mostly male delegates who are in the lead, and who have continuation in the negotiation process as well.' Negotiating skills are built by participating in several COP talks and being permanently involved in the issue. 'Regarding female delegates, I hardly can see continuity. In the last four years that I am attending COP, I hardly saw any female delegates in Bangladesh team who have continuation to attend each year's negotiations,' Zaman continued. 'I believe a well gender balanced negotiation team can prioritize and raise so many crucial issues, especially on equity and women's vulnerability on climate change. But to make a well balanced negotiation team, female delegates' capacity needs to be improved and continuity in the negotiation process needs to be secured.'

Post Bali, there was suddenly a new issue – international NGOs, UN agencies and the World Bank all wanted major climate change programmes in Bangladesh. Each agency, like the UN Food and Agriculture Organization (FAO), wanted its own climate change programme, noted Prof. Nishat. So the government forced interministerial cooperation to prevent competition and insisted that donor agencies follow the BCCSAP. 'We had a particularly big fight with the World Bank.' Nishat noted. These battles continue, as we explain in Chapter 11. Agencies continue to push pet projects, often at the same time promoting the rhetoric of coordination and harmonization as a way of donors and international agencies keeping control and profiling their 'brand'.

Bangladesh has also played a role in helping the UN's Least Developed Countries (LDCs) group to be more organized and more coherent, in part through various meetings that are not intended as formal negotiating bodies but rather as ways to work out issues and positions independent of the industrialized countries. Thus Bangladesh was the third chair (2011–13) of the Climate Vulnerable Forum and it was a founder of the Cartagena Dialogue for Progressive Action in 2010 after the collapse of the Copenhagen COP 15 in 2009. Quamrul Chowdhury was a lead negotiator on finance for the LDCs group. At COP 17 at Durban an Ad Hoc Working Group on the Durban Platform for Enhanced Action was set up to report to COP 21, and Ziaul Haque, deputy director, Ministry of Environment and Forests, was head of 'workstream 2' on mitigation and adaptation before 2020.

Bangladesh has also been pushing for the LDCs and G77 to have common positions based on expertise and clarity, to strengthen their statements. Much of this expertise comes from Bangladesh, which was one of the first LDCs to have an expert team, and is still one of the most powerful, along with Sudan, Gambia and Nepal.

In 2015 UNEP named Prime Minister Sheikh Hasina a Champion of the Earth 'for outstanding leadership on the frontline of climate change'. UNEP Executive Director Achim Steiner said, 'Bangladesh has placed confronting the challenge of climate change at the core of its development.' The award cites the BCCSAP and Bangladesh as the first country to set up its own Climate Change Trust Fund, in 2010. UNEP recognized that 'the government currently earmarks 6–7 per cent of its annual budget – some US$ 1 billion – on climate change adaptation, with only 25 per cent of this coming from international donors.'[9]

---

9 'Bangladesh Prime Minister Wins Top United Nations Environmental Prize for Policy Leadership', UNEP press release, 14 September 2015. Accessed 8 September 2016, http://web.unep.org/champions/multimedia/news/bangladesh-prime-minister-wins-top-united-nations-environmental-prize-policy-leadership.

Sheikh Hasina and the Awami League were re-elected in January 2014. But domestic political machinations temporarily weakened Bangladesh's global position. A new Environment minister was named – Anwar Hossain Manju, chair of Jatiya Party (JP-Manju), a small party in coalition with the Awami League. He had wanted a bigger ministry and was not pleased with Environment; furthermore, he had made public statements that negotiations were a waste of time and money. Bangladesh was always unusual in having such a large non-government component in its official delegation, and shortly before the Lima COP 20, Manju reversed this, limiting the delegation to only government officials. All the COP meetings have a huge media and NGO presence, but only delegates with a government badge can go into the actual negotiations. So at Lima in 2014, 'Bangladesh lost its voice in negotiations,' commented one of the excluded non-government people. 'Government officials come and go, typically attending just one or two COPs, but the non-government side is continuous. It is an intellectual resource and the institutional memory,' Saleemul Huq told us.

The national delegation may have been weakened, but by then Bangladeshis were embedded elsewhere in the negotiation process. One of the younger generation, Hafijul Islam Khan, is a part of the ten-member LDC Group Core Team, and has been responsible for Loss and Damage since 2011. Part of Saleemul Huq's ICCCAD, he had been a member of the Bangladesh COP team until the new minister changed the policy. Gambia has been trying to raise its profile and had chaired the LDC group 2011–12, so it took Khan onto its national team for the Lima COP.

The civil society protests were heard, and for the Paris COP 21 in 2015, Nurul Quadir, joint secretary at the Ministry of Environment and Forests, organized the team. He returned to the mixed team, which incorporated eight independent experts, including Ainun Nishat and Hafijul Islam Khan, and Bangladesh was able to take its lead role again.

Bangladeshis have also become international representatives. Nurul Quadir is a member of the Executive Committee of the Warsaw International Mechanism for Loss and Damage. Q. K. Ahmad is a member of the Executive Board of Clean Development Mechanism of the Kyoto Protocol. Both represent LDCs.[10]

Low-Profile Leadership on Loss and Damage

The Copenhagen COP 15 in 2009 was the first COP where Bangladesh played a very active role. The failure of COP 15 came as a shock to the Bangladesh team and many from the less developed countries. The US, China and India refused to accept the big cuts in greenhouse gas emissions that the developing countries saw as essential. 'Copenhagen failed because it was too ambitious. It was all or nothing, and we got nothing,' argued Saleemul Huq. The response of the Bangladeshi team was controversial – to move to an incremental approach, making small gains one at a time. This included diplomatic efforts to build informal groups of progressive countries, and it involved a much more

---

10 Non-Annex-1 countries.

Box 3.1 What the jargon means

Climate change has its own language. Four terms are important and have specific definitions.

MITIGATION: Climate change is caused by greenhouse gases emitted into the atmosphere. They remain for decades and cause global warming. Mitigation means reducing the emission of greenhouse gases. In the 1990s mitigation was main goal of the annual COP talks – to try and prevent climate change.

ADAPTION: By the end of the 1990s it was clear that greenhouse gases remained in the atmosphere for a long time and that mitigation would not be enough; further it was becoming more obvious that the industrialized countries were not prepared to cut emissions sufficiently. Thus temperatures will rise, sea level will rise, there will be more rainfall and so on. From the early 2000s negotiations increasingly included adaption, which are the changes that have to made to live with climate change – houses on stilts, stronger dykes and cyclone shelters, different crop varieties and so forth.

LOSS AND DAMAGE: By the 2000s it was becoming obvious that there are limits to adaptation. The IPCC in its *Fifth Assessment* in 2014 warned that 'Under any plausible scenario for mitigation and adaptation, some degree of risk from residual damages is unavoidable.'[11] Extreme weather events, floods and temperature rise are all cited by the IPCC.[12] Sea level rise will flood farms and homes and make them uninhabitable, and no amount of adaptation will prevent that. Ever stronger cyclones will destroy even stronger houses. Thus damage is the impacts of climate change that are beyond the limits of adaptation, but can still be repaired, such as rebuilding a destroyed house. Loss is a permanent negative impact of climate change which cannot be reversed and cannot be prevented by adaptation, such as farms lost to sea level rise. The line between the two is often not clear – building a higher dyke to protect the farm is adaptation; if the dyke breaks and floods the farm it is damage, and when the dyke cannot be built any higher and the farm is permanently flooded, it is loss. Loss and damage are usually treated together.

---

intense involvement in the permanent negotiations leading up to the annual COP meetings. The developing country proposals tabled at Copenhagen continued to be the goal. Bangladeshi negotiators stressed three issues:

---

11 IPCC, *Climate Change 2014: Impacts, Adaptation, and Vulnerability. Part A: Global and Sectoral Aspects. Contribution of Working Group II to the Fifth Assessment Report of the Intergovernmental Panel on Climate Change*, (Cambridge, UK and New York: Cambridge University Press, 2014), 1045.

12 K. Van der Geest and K. Warner, *What the IPCC 5th Assessment Report Has To Say about Loss and Damage*, UNU-EHS Working Paper, No. 21 (Bonn: United Nations University Institute of Environment and Human Security, 2015).

- Global temperature rise to be limited to 1.5°C.
- Industrialized countries must accept their responsibility for climate change so far, and must pay the costs of damage and for adaptation. This led Bangladesh to develop and push the concept of 'Loss and Damage.'[13]
- Agreements must be legally binding, not voluntary.

Over the next five years, Bangladeshis have kept 1.5°C and Loss and Damage in the COP documents – despite fierce opposition from the United States – but only as goals and not as commitments. Is that a success and worth all the effort? On one hand, the COP meetings codify the global consensus – a lowest common denominator but one that has been gradually rising and has led to real promises of cuts in greenhouse gases. It also led to increases in funding. On the other hand, as we noted in Chapter 2, the emission cuts promised for the Paris COP 21 in 2015 will lead to a temperature rise of over 3"C in the next century – double of what is seen as acceptable, and causing real problems for Bangladesh. Negotiators from LDCs face the permanent dilemma underlined by COPs 15 and 21: push too hard and the big emitters walk out of the room; push less hard and only be offered slightly less disastrous climate change.

The LDCs are both the least responsible for and most vulnerable to climate change impacts. The 1992 Rio Earth Summit recognized the concept of 'liability and compensation associated with climate change impacts in vulnerable countries'[11], although since then the industrialized countries dug in their heels to prevent anything that might recognize liability and thus require compensation. Pointing to its own experience of sea level rise and worsening cyclones, Bangladesh began promoting the concept of Loss and Damage as an alternative to liability and compensation. There is also an understanding that the costs of loss can be substantial – when land becomes uninhabitable, the climate change refugees need social protection measures and a range of support for resettlement and retraining.

But Hafijul Khan, the Bangladeshi LDC negotiator, is clear: 'We want mitigation – reduction of emissions to curb climate change. We don't want adaptation and compensation for climate change – we don't want climate change. But Loss and Damage is important to remind the industrialized world that if they don't mitigate, they will have to compensate.'

The phrase Loss and Damage first appeared in the 2008 COP 13 Bali Action Plan, which was the first COP attended by the more powerful Bangladeshi team organized by C. S. Karim. Since then, Bangladesh has taken the lead on Loss and Damage, but after the collapse of COP 15 in Copenhagen, Bangladesh has been following its incremental strategy trying to slowly push the issue up the agenda. COP 16 in Cancun in 2010 agreed to study Loss and Damage. COP 17 in Durban in 2011 and COP 18 in 2012 in Doha both agreed there should be an international mechanism for Loss and Damage.

---

13 Saleemul Huq, Erin Roberts and Adrian Fenton, 'Loss and Damage', *Nature Climate Change* 3 (2013): 947.

14 Erin Roberts, *Loss and Damage in Vulnerable Countries Initiative: Bangladesh Leading the Way on Loss and Damage* (London: Climate and Development Knowledge Network, 2012).

The breakthrough was at the Warsaw COP 19 in 2013. The G77 and China pushed hard for a formal establishment of the mechanism agreed in principle at Durban, while the OECD countries simply wanted to continue with a vague work programme and were increasingly intransigent. The LDCs walked out at one point and there was fear of another Copenhagen. The US decided it needed an agreement and opened 'back channel' negotiations, involving key Bangladeshis. The US finally said it would accept a formal mechanism so long as it did not include liability or compensation. When ministers finally agreed a deal at 4.30 am, it included the Warsaw International Mechanism for Loss and Damage.[15] An interim executive committee was established, which was made permanent at the Lima COP 20 in 2014. The Warsaw mechanism accepts the need 'to address loss and damage' but is careful never to suggest liability or compensation.

The Paris COP 21 adopted a two-part document, a 'Decision' that accepts the Paris Agreement as an annex.[16] The Paris Agreement 'recognize[s] the importance' of Loss and Damage and makes the Warsaw Mechanism a formal part of COP. But the agreement avoids any suggestion of liability or compensation, saying only: 'Parties should enhance understanding, action and support, including through the Warsaw International Mechanism, as appropriate, on a cooperative and facilitative basis with respect to loss and damage.[17] The Decision goes further and explicitly says, 'Article 8 of the Agreement does not involve or provide a basis for any liability or compensation.'

Baby Steps When Giant Strides Are Needed

The international negotiation processes and UN meetings around climate change can seem like a lot of hot-air – everyone talks and then nothing happens. However long-winded and cumbersome these processes seem, sometimes they can genuinely 'change' the world. An outstanding example is the first climate change agreement, the Montreal Protocol of 1987 on limiting damage to the stratospheric ozone layer. This led to the total phasing out of chloroflurocarbons from refrigerators and air conditioners, once a major contributor to global warming and the equivalent of reducing CO_2 emissions by 135 billion tonnes until now.[18] The ozone layer is expected to be reconstructed by 2060, and the US Environmental Protection Agency estimates that at a global level, up to 2 million cases of skin cancer may be prevented each year by 2030. So, climate change negotiations can deliver.[19]

---

15 http://unfccc.int/adaptation/workstreams/loss_and_damage/items/8134.php, Accessed 25 January 2016.

16 Framework Convention on Climate Change, 'Decisions Adopted by the Conference of the Parties' (adopted at COP21, 12 December 2015), FCCC/CP/2015/10/Add.1, 29 January 2016.

17 Paris Agreement Article 8.

18 'Montreal Protocol Parties Devise Way Forward to Protect Climate Ahead of Paris COP21', United Nations Environment Programme, 6 November 2015, Accessed 18 April 2016, http://www.unep.org/newscentre/Default.aspx?DocumentID=26854&ArticleID=35543.

19 'Collated Research Reveals Full Scale of Montreal Protocol's Ozone Layer Repair Work', United Nations Environment Programme, 4 November 2015, accessed 18 April 2016, http://www.unep.org/NEWSCENTRE/default.aspx?DocumentID=26854&ArticleID=35539.

'The Paris Agreement is a genuine triumph of international diplomacy,' said Prof Kevin Anderson, deputy director of the Tyndall Centre for Climate Change Research, University of Manchester, because by recognizing that climate change is real and mitigation is required, it means that 'the sceptics have lost the argument.'[20] The Decision goes further by 'emphasizing with serious concern the urgent need to address the significant gap' between the voluntary pledges and what is needed to hold the increase to 'well below 2°C above pre-industrial levels.'

The Agreement is also important because for the first time it recognizes four points pushed by the developing countries: 1) included the concept of climate justice, albeit in a very grudging way, by noting 'the importance for some of the concept of "climate justice"' 2) recognized 'equity and the principle of common but differentiated responsibilities and respective capabilities' so that already industrialized countries must do more and move faster 3) accepted that actions must be taken 'in accordance with best available science' and 4) put stress on the need for 'transparency' in the mitigation promises and actions of the industrialized countries.

But the recognition of the problem is accompanied by no binding commitment to do anything. 'The incremental approach to climate change has fundamentally failed to deliver,' argues Anderson. Countries made voluntary pledges to reduce emissions but these are not binding and cannot be enforced or effectively monitored. Furthermore, despite the 'serious concern' about the 'significant' shortfall in the voluntary pledges, governments do not have to do anything more for five years, and those five years of extra emissions mean that even greater cuts will be needed in 2020. The need to move away from fossil fuels is not even mentioned in the Paris Agreement.

Perhaps the biggest disappointment for Bangladesh was the lack of funding. All it could gain from the Decision was that it 'strongly urges developed country Parties to scale up their level of financial support, with a concrete road map to achieve the goal of jointly providing $100 billion annually by 2020 for mitigation and adaptation' in developing countries.

But United Nations climate chief Christiana Figueres has used International Energy Agency figures to suggest that it will take ten times that amount, $1 trillion per year, to achieve the shift from a coal- and oil-based economy to the cleaner fuels and technologies that would help keep warming below the dangerous threshold of 2°C.[21] The World Future Council estimates it would take double that – $2 trillion – to get down to 1.5°C.[22] Although this is 10 or 20 times what the Paris Agreement proposes, it is still far less than current fuel subsidies cost. An IMF study estimated that global post-tax energy subsidies

20 Kevin Anderson, 'Going Beyond "Dangerous" Climate Change', public lecture at London School of Economics, 4 February 2016.

21 Suzanne Goldenberg, 'UN Climate Chief Calls for Tripling of Clean Energy Investment', *Guardian* (London) 14 January 2014, http://www.theguardian.com/environment/2014/jan/14/un-climate-chief-tripling-clean-energy-investment-christiana-figueres.

22 Matthias Kroll, 'We Print Money to Bail out Banks. Why Can't We Do It to Solve Climate Change?', *Guardian* (London) 30 January 2016, http://www.theguardian.com/global-development-professionals-network/2016/jan/30/print-money-climate-change-green-bond-quantitative-easing?

rose from $4.2 trillion in 2011 to $5.3 trillion (6.5 per cent of global GDP) in 2015.[23] For comparison, the amount of money printed by central banks after the 2008 economic crisis to bail out banks and stimulate the global economy, known as 'quantitative easing', was about $5 trillion.[24]

Conclusion: Bangladesh in the Lead, but Can Small Victories Halt the Rising Sea?

In Paris the developing countries continued to push the Copenhagen agenda, and gained a small victory with the inclusion of an agreement 'to pursue efforts to limit the temperature increase to 1.5°C above pre-industrial levels'. Although the official goal is still 'below 2°C', this is the most prominent statement so far on 1.5°C. Five COPs after Copenhagen, there is better coordination and cooperation between the developing countries, and a slow and grudging change of attitudes by the industrialized countries and India and China, leading to promises to reduce greenhouse gases more than had been offered before. But this is not enough, and the interminable negotiations have proved a very hard slog.

Bangladeshi scientists recognized the problems of climate change very early. They saw the impact it would have on their low-lying delta and they became important participants in the international campaign. In the past decade, Bangladesh has used its expertise to play a leading role on the LDC side in the hard COP negotiations with the industrialized and industrializing counties. Bangladesh can claim substantial credit for promises to cut greenhouse gases – even if not enough – and for the very reluctant recognition by industrialized countries that they might be responsible for, and perhaps liable for, the loss and damage caused to the climate by two centuries of industrialization. In global forums, Bangladesh has won at least some breathing space, gaining some additional time to adapt to climate change and to perhaps shame the industrialized nations into more action. Far from being a helpless victim, Bangladesh is fighting successfully to moderate climate change. But a huge battle remains.

---

23 David Coady et al., 'How Large Are Global Energy Subsidies?' IMF Working Paper WP/15/105, 2015. Accessed 8 September 2016, http://www.imf.org/external/pubs/ft/wp/2015/wp15105.pdf.
24 'What Is Quantitative Easing?', BBC, 3 December 2015, http://www.bbc.co.uk/news/business-15198789.

Chapter Four

SEA LEVEL RISE AND THE VULNERABLE COAST – WHERE FARMERS KNOW MORE THAN ENGINEERS

As the motorcade of the local member of parliament, Abdul Wahab, passed through Kalishakul village, it was attacked by several thousand angry villagers. Twelve vehicles were burned. Wahab escaped, but a water board engineer, the sub-district chairman and a police superintendent were injured in the 2 June 2012 incident.[1] The motorcade was on its way to launch construction of a dam in Kapila Beel, which many local people opposed. It was one of the most violent incidents in a 50-year struggle of local people to reverse a system of dykes and polders[2] built in the 1960s and 1970s, which caused thousands of hectares to become permanently flooded and waterlogged, with hugely detrimental effects on local agriculture and livelihoods. Local people want an updated version of a system that dates back at least to the seventeenth century and involves using the silt deposited by annual flooding to raise the level of the land. When people launched their campaign, they had never heard of climate change, but they have won broad support from leading political and technical figures who realize it is an important response to rising sea levels.

Less than a year later, on 14 March 2013, the people of Chalan Beel formed a 220 km human chain along the banks of the dying Boral River. 'We demand removal of all sluice gates and cross dams on the river,' said Dr Matin, the spokesperson for the four groups that organized the demonstration.[3] In both cases the demand was for changes to structures – dams and embankments – and to combine ancient knowledge with modern technology to work with nature and not against it. In both cases the problem is large earthworks imposed by engineers and international agencies, without taking into account the very complex structure of the Bengal delta and without paying attention to local knowledge.

---

1 'Two Police Officials Withdrawn over Attack on Wahab's Motorcade' and '5000 Sued in Two Cases for Attack on Whip's motorcade', *Financial Express*, Dhaka, 3 and 4 June 2012; '2 OCs Closed', *Daily Star*, Dhaka, 4 June 2012.

2 Although 'polder' comes from the Dutch and originally meant low-lying land reclaimed from the sea, the word is now used more broadly to mean low-lying land enclosed by dykes that is sometimes below the water level of surrounding water bodies.

3 Ahmed Humayun Kabir Topu, 'Save Boral to Save Chalan Beel', *Daily Star*, Dhaka, 15 March 2013.

For centuries Bengalis have been building embankments and canals to try to manage this ever-changing delta, to try to increase agricultural production while controlling floods and defending against cyclones. But not all of these interventions have been successful. In particular, some embankments and dams built in the colonial era and after independence disrupted water flows in ways that did considerable harm. The past 50 years has seen a struggle – on one side engineers and state and donor agencies favouring large dams, sluice gates and other major building works; and on the other side local communities who want smaller interventions based on a better understanding of the delta, and who want the most disruptive structures removed. The struggle continues, but the balance is changing.

In this chapter we tell the story of this struggle. First, we explain that although they superficially seem the same, there are huge differences between the coastal areas in the south and those dominated by the rivers further north. This chapter is about the coastal areas, where climate change will hit first, while Chapter 6 looks at the river zones. Second, we point to the natural processes in the coastal zone that raise the land level and make it possible to raise the land to match sea level rise. Third, we report on the engineering era of trying to tame nature. Fourth, we look at the backlash where local communities, often taking direct action such as cutting embankments, have been seen to understand the delta better than foreign consultants and national engineers. This led to a major change in thinking by policy makers, scientists and especially those who do increasingly important computer modelling of Bangladesh's tides and rivers. The result is a concept known as Tidal River Management (TRM). Fifth, we point out that introducing TRM faces social, political, economic and technical challenges and will require ongoing innovation and experimentation. We conclude the chapter by arguing that working with natural processes, rather than against them, may be at the core of adaptation in Bangladesh.

The coastal area is the part of Bangladesh that will be most affected by climate change – especially rising sea level and more fierce cyclones. But it is also the area that has most experience adapting to the difficult environment and holds lessons for the rest of the country. In this chapter we look at using natural processes to raise the land level and in Chapter 5 we show how the remarkable cyclone shelter programme is saving lives.

Climate change hits the coastal zone first, and the impact of climate change and the response of the next few decades will be the model for the rest of the country. More than anywhere else, this is where Bangladeshis must struggle to keep their heads above water. Here and elsewhere they are trying to combine modern technology with ancient knowledge and techniques.

River and Coastal Zones Are Very Different

The delta is very complex and there are huge differences between the north and the south, so we divide the discussion into two chapters. Chalan Beel is near the junction of the Ganges-Padma and Brahmaputra-Jamuna rivers, 130 km northwest of Dhaka, and is discussed in Chapter 6, about river flooding. Kapila Beel is in the coastal district of Khulna, 150 km southwest of Dhaka, and is discussed in this chapter, about tidal flooding.

For a small country, Bangladesh's soils, geography and geological history are very complex, and we present only a very simplified version here. Hugh Brammer's *The Physical Geography of Bangladesh*[4] gives a more complete picture. Much of Bangladesh is a delta that has been built up over millions of years. Although there are no accurate measurements, it is usually estimated that more than one billion tonnes of sediment are carried from the Himalayan mountains each year by the Ganges and Brahmaputra rivers.[5] Parts of the delta slowly sink as the sediment compacts and settles, but each year new sediment is carried downstream, building up the delta. Part of the sediment is dropped by the rivers as they flow through Bangladesh and part carried out to the Bay of Bengal.

A technical point about sediment is important to understanding the shaping of the delta. Faster flowing water carries more sediment. As water slows, it drops the sediment, building up the land; as the water moves faster, it carries more sediment and even scoops up more silt, eroding the river banks and deepening the channels.

Moving across a relatively flat land, these mighty rivers have numerous smaller branches that change course, sometimes abandoning courses and carving out entirely new channels. This creates a huge network of rivers and streams often used for transport. Faster flowing water carries more sediment, but on the shallow edges of rivers and streams, the water is slower and more sediment is dropped, building up natural levees or banks. Beyond these natural levees are shallow bowls known as *beels*, which fill with water during the June-September monsoon and are largely dry by March. At any given time the area flooded by river water is relatively small, but these smaller tributary rivers are constantly changing leaving a pattern of ridges 1–3 m higher than the adjoining floodplains. Villages are built on these natural levees. The major rivers, the Meghna, Ganges and Brahmaputra, do not change courses as often as the smaller ones, but they can erode sections of riverbank, cutting away up to 2 km from the bank in just one monsoon season. In other places, the rivers slow and drop their sediment, widening the banks or creating entirely new islands known as *chars*. In just four months, land levels can be raised by 2 m or more. As noted in Chapter 1, each year the floods are different. Chalan Beel is in the river-flooding zone and is discussed in Chapter 6. Climate change will bring more intense river and monsoon floods, and we consider the response in Chapter 6.

Whereas the river floodplains have been created by sediment carried down by the rivers in the monsoon, in the coastal zone, just the opposite happens, and land is built up by the tides during the dry season. We noted that part of the sediment carried into Bangladesh is dropped by the rivers, while huge amounts of sediment are carried out into the Meghna estuary and to the Bay of Bengal, creating new land along the coasts and small and large islands. The huge monsoon flows stop seawater coming upstream, but during the dry season high tides reach far into the country. The sea near the coast is muddy, and each high tide drops a small amount of silt that builds up over time. In parts of the coast, islands and sandbars are created, which are often exposed at low tide and covered at high tide and have been colonized by mangroves and other plants to create the Sundarbans. This

---

4   Hugh Brammer, *The Physical Geography of Bangladesh* (Dhaka: University Press, 2012).
5   Mohammad Rezwanul Islam et al., 'The Ganges and Brahmaputra Rivers In Bangladesh: Basin Denudation and Sedimentation', *Hydrological Processes* 13 (1999): 2907.

coastal area is always shifting. Analysis of satellite images shows that between 1972 and 2010 Bangladesh lost 1121 km² of land to erosion, but gained 1885 km² of new land along coasts, islands and in the Sundarbans as sediment was deposited.[6] So each year the mouth of this hugely dynamic delta creates 50 km² and destroys 30 km² – a net gain of 20 km² a year. Land tends to be eroded from the north side of islands and added on the south side, where it is initially unstable and less compact but is bonded over two decades by mangroves.[7] The coastal zone in the southwest includes the village of Kalishakul and the country's third largest city, Khulna, while the coastal southeast includes the second largest city, Chittagong. We discuss the coastal zone in this chapter. Here the climate change issues are cyclone and sea level rise, and adaptations will be essential. But in many places in coastal areas it will be possible to raise the land levels to keep up with sea level rise.

As well as the coastal and river zones there is an intermediate estuarine zone. After the Ganges and Brahmaputra merge and join the Meghna, the river is 20 km wide, and water flows down hundreds of other smaller channels. This huge amount of fresh water pouring down the rivers and into the Bay keeps the salty sea water out of most of the county. However the floods only occur during the monsoon season, June through September. In the dry season high tides reach more than 300 km upstream. Thus many channels that carry huge amounts of fresh flood water during the monsoon become tidal channels in the dry season. Dhaka sits on the boundary between the river and estuary zones, affected both by monsoon river floods and dry season tides.[8]

Across the country there are basins and beels that serve two important hydrological functions – they store water during the monsoon, which reduces the downstream flow and erosion at peak flood periods, and the water recharges the aquifers so that groundwater can be used for irrigation in the dry season. Over centuries, tall and rapidly growing rice varieties were developed; they are planted at the start of the monsoon and grow fast enough to keep their heads above water as the beel fills. Houses are built on levees or constructed mounds to be above flood levels.

Normal floods cover a significant part of the country and have long been recognized as restoring the fertility of the land. It was always assumed that in some way the silt brought the fertility, but folk wisdom is not always precisely correct. In fact the silt does not contain nitrogen, although it contains essential minerals that are released over time by weathering. Instead the fertility comes from blue-green algae (cyanobacteria) that grows on soil, in the water and on plant stems in the flooded land and fixes nitrogen and became a natural fertilizer.[9] Weeds and crop residues also decay in the flood water, adding nitrogen. Finally, sometimes there is a very small amount of organic material mixed with the sediment.

---

6   Md. Asraful Island, Abdul Baquee Khan Majlis and Md. Bazlar Rashid, 'Changing Face of Bangladesh Coast', *Journal of NAOMI* 28 (2011): 16.

7   Md. Golam Mahabub Sarwar and Aminul Islam, 'Multi Hazard Vulnerabilities of the Coastal Land of Bangladesh', and Mesbahul Alam et al., 'Coastal Livelihood Adaptation in Changing Climate: Bangladesh Experience of NAPA Priority Project Implementation', in Rajib Shaw, Fuad Mallick and Aminul Islam, eds, *Climate Change Adaptation Actions in Bangladesh* (Tokyo: Springer, 2013), 126, 133, 256.

8   Dhaka is on what is known as an 'uplifted block'; see Brammer, '*The Physical Geography*', 30–31.

9   Brammer, '*The Physical Geography*', 102–3.

Because much of the country appears to be composed of beels and embankments, and all of these beels fill with rainwater during the monsoon, outsiders have often missed an essential difference. More northern beels are affected by monsoon season river flooding and coastal beels affected by dry season tidal flooding, which creates different ecologies. Understanding that beels in the northern river-influenced zones are totally different from those in the coastal zone is essential for any intervention, including responding to two different impacts of climate change – heavier monsoon rain and worse floods, on one hand, and sea level rise, on the other.

Box 4.1 Ancient kilns show land can match sea level rise

Excavations of ancient salt kilns by German archaeologists have shown how the sedimentation process not only compensates for the land compacting and sinking, but also for sea level rise. Buried in mud under mangroves and rice fields at the edge of the Sundarbans, the team found the remains of kilns that had been used to produce salt on an industrial scale. Historic records showed they had been built at the spring (dry season) high tide level and tests showed they were last used around the year 1700.[10]

Cyclones played a role. The archaeologists suggest that production stopped after the devastating cyclone of 1699, and the kilns were exposed by erosion caused by Cyclone Sidr in 2007. Seven kilns were found, one group at 70 cm and another group at 150 cm below current spring high tide level. IPCC estimates that sea level has risen by about 20 cm between 1900 and 2015.[11]

Yet above these sunken kilns are mangroves and rice fields – above spring high tide level. Thus the land is staying above sea level. Similarly, a detailed study published in 2015 in the prestigious *Nature Climate Change*[12] included precise measurements of sea and land levels on the edge of the Sundarbans, and it found that despite general sea level rise, the land was at the same level with respect to the sea as in the 1960s. Because the sea level is obviously rising, the sediment naturally deposited in this area has raised the land more than enough to compensate for both subsidence and sea level rise. Will sediment drops continue to raise the land to keep up with sea level rise caused by global warming?

---

10 Till J. J. Hanebuth et al., 'Rapid Coastal Subsidence in the Central Ganges-Brahmaputra Delta (Bangladesh) since the 17th Century Deduced from Submerged Salt-Producing Kilns', *Geology* 41 (2013): 987–90.

11 If the IPCC estimated sea level rise is subtracted, this means sinking of the kilns of 50 cm and 130 cm in 300 years, which means that sinking, largely due to sediment compacting, has been between 2 and 4 mm per year, which is similar to rates proposed elsewhere. See Hugh Brammer, *Climate Change, Sea-level Rise and Development in Bangladesh* (Dhaka: The University Press, 2014), 154–55.

12 L. W. Auerbach et al., 'Flood Risk of Natural and Embanked Landscapes on the Ganges–Brahmaputra Tidal Delta Plain', *Nature Climate Change* 5 (2015): 153–57 and 492–93. doi: 10.1038/NCLIMATE2472

Traditional agriculture in the coastal areas was to grow *aman* rice in the rainwater-flooded beels during the monsoon. Monsoon rains and river water help to keep the sea water away. However, during the dry season the tide reaches further inland, entering and spreading out across the beels. We noted earlier that faster flowing water carried more sediment, and as the water slows, the sediment is dropped in the beels, raising the level of the beel to compensate for the slow lowering of the delta as the sediment compacts. Twice a day, as the tide goes down, the water retreats to the channels and moves faster, scouring out the channels and keeping them deep enough for boats. So this natural two-part process raises the level of the farmland, while keeping the water courses open. The mud from the sea is salty, and the final piece of the jigsaw is that the heavy monsoon rains wash the salt out of the soil each year; in June, the first month of the monsoon, soil salinity is reduced by three-quarters, and remains low until February.[13]

The first engineering intervention in the coastal zone was made during the time of the Mughals, who developed a system called 'eight-month embankments' (*osthmeshe bundh, ayshtomaisha badh*) to make this natural system more effective. Temporary low earth structures with wooden sluices were built each year to keep fresh rainwater in the beels for growing rice and to keep the sea water out. At the end of the season these low embankments were breached or wooden sluices were opened from January to April to allow the sea water in and to drop the sediment. Various drainage and irrigation channels were built.

Thus in many areas it is possible in traditional systems to grow a rice crop using rain and flood water during the monsoon, and carrying out other activities during the dry season, including fishing and raising salt-water shrimp. This is the indigenous knowledge of Bangladeshi farmers, developed over centuries. Indigenous knowledge is often very local; like many things in Bangladesh, salinity varies significantly across the coastal zone, in terms of location, amount and seasonality, and is affected by rainfall and flooding patterns. The farmers knew this, but the engineers were not listening.

The southwest is a coastal area that has 28 per cent of Bangladesh's population in 32 per cent of the country's area. This whole region is low, mostly less than 3 m above sea level. As one moves south, the land breaks up into innumerable tidal rivers and canals surrounding islands; boats become the main means of transport. Drainage channels cut across the islands. When we visited, we had to leave our car at a ferry pier. On the other side we took motorcycle taxis to the next, smaller ferry, and then other motorcycles across another island. At the southern end are the Sundarbans, a network of mudflats and tidal wetlands, which is the largest contiguous mangrove forest in the world, and which is, perhaps most famously, home of the Bengal tiger.

Many of these islands are surrounded by dykes built in the 1970s, making them Dutch-style polders. Brick roads run across the tops of the dykes – many almost covered by drying rice. Indeed, all life seems perched on tops of the dykes. Increasingly, flooding and waterlogging has forced people to move their houses up onto the dykes. Every bit of space is used – the roofs of houses are shared by pumpkin vines and solar panels. The

---

13 D. Clarke et al., 'Projections of On-Farm Salinity in Coastal Bangladesh', *Environmental Science Processes & Impacts* 17 (2015): 1128.

fields below are filled with rice, sugar cane, vegetables and shrimp and fish ponds. But flooding and waterlogging and recently repaired breaches in the dykes showed all was not well.

The US, Krug, Pakistan and the Engineering Years

A chain of engineering interventions started after the 1947 partition from India, when the new United Nations looked for ways to help the newly independent Pakistan. Many of them proved misguided. The cold war, engineers, neo-liberalism, well-meaning outsiders and a heavy dose of domestic politics all contributed to how the land was reshaped in ways that ignored both local wisdom and reality of Bangladesh as an active delta. Sixty years later, many of these interventions are slowly being changed. But global politics played a role in events leading up to the demonstration in Kalishakul that opened this chapter.

In Pakistan's early years, the top priority was to push for a jump in food production, first through irrigation and then through 'Green Revolution' high-yielding rice varieties. In East Pakistan this meant increasing monsoon season aman rice through more surface water irrigation.

A Dutch water and irrigation engineer, Willem Johan van Blommestein,[11] became part of the UN Food and Agriculture Organization (FAO) team in East Pakistan in 1951. He designed the Ganges-Kobadak Project, a massive irrigation system with a large water intake on the Ganges and a main canal to transport water by gravity 73 km south and irrigate 142,000 ha in parts of the estuary and coastal zones. Construction started in 1954 and the first phase was completed in 1970.

Unfortunately, the Bangladesh delta is different from the Netherlands and the Ganges-Kobadak Project had huge problems. The main intake channels to the large pumps quickly silted up or were blocked by water hyacinth; variable electricity supplies burned out the pump motors. The green revolution rice variety introduced, IR5, was a dwarf variety and could not withstand the normal levels of flooding, and took too long to grow, so farmers could only get one crop per year instead of two. Irrigation water was not available at the right time or polders flooded and could not be used. In particular, the new embankments often cut off pre-existing small channels and canals used by the farmers. This led farmers in the mid-1960s to cut openings in the embankments. Hugh Brammer recalls when he first arrived in 1961 in the then East Pakistan for the FAO that 'the newly built embankment along the Ganges River [...] was cut in more than 90 places'.[15] Farmers said they had not been consulted about the Ganges-Kobadak Project and felt that the embankment would prevent the river from flooding their land which would stop the river contributing the water and fertility they were accustomed to. In

---

14 W. Ravesteijn, 'Blommestein, Willem Johan van (1905–1985)', in *Biografisch Woordenboek van Nederland.* Accessed 18 April 2016, http://resources.huygens.knaw.nl/bwn1880-2000/lemmata/bwn5/blommestein.

15 Hugh Brammer, *Can Bangladesh Be Protected from Floods?* (Dhaka: The University Press, 2004), 113fn.

a 1972 evaluation for USAID, John W. Thomas said that to protect the embankments police fired on the peasants and killed several.[16]

The next attempt at engineering followed serious, albeit not extraordinary, floods three years in a row, in 1954, 1955 and 1956, which drew global attention to East Pakistan. In Chapter 1 we noted the importance of Pakistan to the US in the cold war; in the 1950s nearly all aid to Pakistan was from the US. The resident representative of the UN Technical Assistance Board, Huntington Gilchrist, was from the US and publicly expressed support for 'General Ayub Khan's dictatorship' and his support for the West.[17] Gilchrist was on first name terms with Central Intelligence Agency Director Allen Dulles and sent him personal reports from his travels.[18] Gilchrist personally organized a UN mission headed by Julius Krug, who had been US Secretary of the Interior,[19] 1946–49. More importantly in 1938–41 Krug had been chief power engineer of the Tennessee Valley Authority (TVA), a giant government-owned company responsible for everything from flood control, electricity generation and economic development to malaria control and agricultural extension in the Tennessee River valley.

Krug included in his 1956–57 mission van Blommestein, a former TVA director of flood control, and two British former colonial civil servants. Perhaps reflecting Krug's own background, the mission proposed 'the establishment of a Government Corporation [...] which would have complete authority and responsibility for dealing with all the related aspects of water and power development.' Such a corporation should be supported by 'one or more of the world renowned engineering and management firms.'[20] This also was seen as something similar to the Dutch waterways agency.[21] The idea was supported by aid agencies that wanted to bypass a government which some saw as nationalist, corrupt and inefficient.[22]

The Krug mission proposed a Dutch-style system of embankments that would enclose polders to control damaging floods and to regulate water in a way that allowed peasants to grow two crops per year of high-yielding varieties of rice. There would be

---

16 J. W. Thomas, '*Development Institutions, Projects and Aid in the Water Development Program of East Pakistan*' (Washington, DC: Agency for International Development, 1972), 15, 16, 18, 20.

17 Huntington Gilchrist, 'Technical Assistance from the United Nations – as Seen in Pakistan', *International Organization* 13 (1959): 506.

18 Correspondence between Allen Dulles and Huntington Gilchrist, no date and 30 January 1959, released by the CIA. Accessed 8 September 2016, http://www.foia.cia.gov/sites/default/files/document_conversions/5829/CIA-RDP80R01731R000200110062-4.pdf.

19 Not to be confused with "Interior" in most other countries, which usually includes police. In the US the Secretary for Interior is a cabinet member who is responsible for national parks, the US Geological Survey, some mining issues and relations with Native Americans.

20 United Nations Water Control Mission, *Water and Power Development in East Pakistan* (New York: United Nations Technical Assistance Programme, submitted 31 May 1957, published 1959), report ST/TAO/K/PAKISTAN/2, vi, vii. Known as the 'Krug Mission'.

21 Netherlands Water Partnership, 'Bangladesh – The Netherlands: 50 Years of Water Cooperation,' (The Hague: Netherlands Water Partnership for the Dutch Government, no date but probably 2014), 6.

22 Thomas, 'Development Institutions', 2.

'large-scale' dry season diesel and electric pumping of water for irrigation. Support was expressed for the ongoing Ganges-Kobadak Project and suggested that the project's system of 'large polders which are enclosed by ring embankment that prevent ingress of salt water' should be extended to other saline areas.[23]

In 1959 the military government created the East Pakistan Water and Power Development Authority (WAPDA), as proposed by Krug, and the new US Agency for International Development (USAID) began to fund two major projects. The first was to create a master plan. By 1966, there were 85 foreign staff from international engineering companies working for the WAPDA. In 1968, after nine years of work, the contract of the US consulting company was terminated and a new team of 38 Dutch and Canadian consultants was hired to work in WAPDA to try to develop the master plan.[24] Even at the time, questions were being raised about the master plan. In 1966 the World Bank advised against it and Thomas in his 1972 report emphasized small flood control projects and the expansion of irrigation by means of low-lift pumps and tube wells.[25] Initially ignored, these ideas came to fruition with the successful introduction of irrigated, dry season *boro* rice two decades later (see Chapter 7).

The other USAID-funded project was the Coastal Embankment Project, which in the 1960s through 1980s built nearly 6,000 km of embankments enclosing 139 polders, some as far as 150 km inland from the sea.[26] These were south and west of the Ganges-Kobadak Project. Nature was to be tamed and the coastal region re-engineered to Dutch-style polders using European and US experience, despite this being a very different kind of delta. Indeed, Bangladeshis call the enclosed areas 'polders'.

By 1970 almost $2 bn in aid money had been allocated to WAPDA, of which 15–20 per cent had been paid to firms in donor countries to hire consultants.[27] Some projects involving large dykes and high-capacity pumping systems were started. The designers simply adapted developed-world models and did not understand that this delta was different from those in the Netherlands or the Mississippi and Tennessee River valleys of the US.

In a 1972 evaluation for USAID, John W. Thomas said that the WAPDA was 'fundamentally an engineering organization' and that it and its consultants had 'only superficial knowledge of the environment'. He pointed to 'the gross neglect of the agricultural dimension' and noted that 'farmers were not consulted nor even informed'.[28] Thomas continued that the 'standard pattern involved aid-financed foreign consultants who designed projects according to their own and aid donors' concepts and standards of efficient design. This meant high cost and long construction

23 Krug Mission, v–vii, xii, 51, 52.
24 Thomas, 'Development Institutions', 6, 12, 25.
25 Brammer, *Can Bangladesh be Protected*, 150.
26 Bangladesh Water Development Board (BWDB), *Coastal Embankment Improvement Project Phase-I, Environment and Social Management Framework* (Dhaka: Ministry of Water Resources, 2013), 1, 14.
27 Thomas, 'Development Institutions', 1,2,8.
28 Ibid., 16, 24, 37.

periods but assured aid support for the project. This usually entailed complex technology that must be constructed in an accessible location so that foreign engineers could supervise the work'.[29] Western engineering corporations and their employees found such work very lucrative.

With the independence war in 1971 and independent Bangladesh's turn to the socialist bloc, work initially stopped. Thomas stressed the power of agencies like the WAPDA, which are conduits for foreign aid, of the aid agencies and their personnel, and of the international consulting firms that 'will resist changes to the system'. He continued, 'The new nation of Bangladesh has a unique opportunity to abandon the institutions and administrative system that served East Pakistan so badly.'[30]

Sadly, it did not. The East Pakistan Water and Power Development Authority was simply renamed the Bangladesh Water Development Board (BWDB) – indeed local people still call it WAPDA.[31] Bangladesh's new military ruler, General Zia, turned to the US and World Bank and the building of dykes and polders resumed, alongside the continued ignoring of indigenous knowledge.

Misunderstanding Sediment Flows

The main purpose of the USAID's Coastal Embankment Project polder system was to increase the production of monsoon aman rice, and there were some initial gains, with farmers in some polders producing substantially more rice. But within a decade some beels were several metres below the levels of the channels and no longer drained. They became filled with stagnant water and became waterlogged; when the soil is so filled with water that soil pores become saturated thus displacing air, trees die and farming becomes impossible. By 1990, over 100,000 ha in Khulna, Jessore and Satkhira districts had become waterlogged, and crops could not be grown. By 2011 this had increased to 128,000 ha.[32]

In Khulna district, the area under monsoon aman rice rose from 262,000 ha in 1972 to 429,000 ha in 1992 in what was seen as a success for the project, but had dropped back to 273,000 ha in 2011 due to waterlogging and the move to dry season boro rice. Only in Barisal district was the large increase maintained, from 274,000 ha of aman in 1972 to 414,000 ha in 2011.[33]

---

29 Ibid., 7–9.

30 Ibid., 42, 43.

31 Leendert de Die, *Tidal River Management: Temporary Depoldering To Mitigate Drainage Congestion in the Southwest Delta of Bangladesh* (Wageningen, the Netherlands: Wageningen University, MSc thesis, 2013), 6.

32 Md. Abdul Awal, 'Water Logging in Southwestern Coastal Region of Bangladesh: Local Adaptation and Policy Options', *Science Postprint* 1 (2014). e00038. doi: 10.14340/ spp.2014.12A0001 and Md. Waji Ullah, Executive Director CEGIS, in *People's Plan of Action: Management of Rivers of South-West Coastal Region of Bangladesh* (Dhaka: Uttaran, 2013), iv. Accessed 8 September 2016, http://www.uttaran.net/publications/peoplesplanofaction.pdf.

33 Hugh Brammer, *Climate Change, Sea Level Rise and Development in Bangladesh* (Dhaka: University Press, 2014), 163.

The dyke and polder system was developed in the Netherlands to claim land from the sea, and involved pumping the water out from land lying below river or sea level. International engineers adapted it for Bangladesh to protect existing farmland from floods. But that reflected a total misunderstanding of how the Bangladesh delta functions, where floods are normally beneficial. In the coastal areas during the dry season, if there are no interventions, at high tide the water spreads over the land and slows, dropping its sediment. As the tide falls, the water flows back towards the sea, faster and without its sediment, and scours river channels and keeps them clear.[34] Raising the level of the beels and lowering the level of the channels go together.

Creating polders surrounded by dykes ended these natural processes, and tidal water and sediment could no longer enter the polders and their beels. The sediment being brought in by the tides could no longer be spread over the entire floodplain and so was dropped in the rivers and channels between the embankments. This started to raise the level of the channels, which often slowed the flow, leading to more sediment being dropped. Meanwhile, the subsidence in the beels was not replaced by new sediment.

This had been predicted decades earlier. Prasanta Chandra Mahalanobis, a Bengali professor at Presidency College, Calcutta, and founder of the Indian Statistical Institute, was commissioned by the government of Bengal to report on the serious floods of 1922 and earlier. In his 1927 study he specifically noted the systems used in the United States and said that 'conditions in North Bengal do not however appear to be at all suitable for this method. [...] The inevitable result of the erection of river embankments will however be the raising of the river beds'.[35] Just 40 years later these methods were used, and the result was as Mahalanobis predicted.

Changes over Time

Two changes since independence, taking place at the same time as the dykes and polders were being developed, have altered the context and removed some of the justification of the projects. Perhaps the most important change was the development of new rice varieties by the Bangladesh Rice Research Institute (BRRI), which had been established immediately after independence. Dyke and polder schemes were designed to improve water control of flooded monsoon aman rice. But BRRI's biggest success of the 1980s was the development of dry season boro irrigated rice, which was much more productive and the boro BR28 variety is now the most common (see Chapter 7). Transplanted aman rice is still the most common in the coastal areas,[36] particularly in western coastal areas, because shallow groundwater is saline, which makes pumped irrigation unsuitable. Part

34 Md. Waji Ullah, *TRM Concept and Its Application in Southwestern Delta* (Dhaka: CEGIS, Powerpoint presentation, 2011).

35 P. C. Mahalanobis, *Report on Rainfall and Floods in North Bengal 1870–1922* (Calcutta: Bengal Secretariat Book Depot, 1927), 43.

36 Ministry of Agriculture, *Master Plan for Agricultural Development in the Southern Region of Bangladesh* (Dhaka: Food and Agriculture Organization of the United Nations, 2013), 28, 49, and Brammer, *Climate Change*, 163.

of the justification of the dykes was that the value of increased aman rice production would justify the cost. But boro rice is increasingly being adopted and a reduced share of rice is aman, while total rice production is rising and prices are falling – undermining the main economic reason for the projects.

The second important change was that in 1975 India completed the Farakka Barrage across the Ganges River just 16 km from the Bangladesh border to divert water into the Hooghly River to stop sediment deposition in Calcutta (now Kolkata) harbour. Built without agreement of Bangladesh, this barrage has been hugely controversial and is said to divert up to 40 per cent of the dry season flow of the Ganges, leading to increased sediment build up in rivers in Bangladesh, causing problems in some polder and embankment areas. Before the barrage was built, salt water intruded on the Ganges during the dry season 274 km inland, but after the barrage was built, it intruded 434 km.[37]

Community Action to Develop Tidal River Management

By the 1980s the problems caused by dykes and polders were becoming more obvious. The engineers and the international agencies continued to look for large-scale solutions. Local people increasingly cut the embankments and began to evolve a modern version of the Mughal-era eight-month embankments, which involved a rotating system of temporary tidal basins and eventually came to be called TRM. After more than 30 years, there is still a struggle between the two approaches being played out in part of the tidal southwest, west of Khulna (Bangladesh's third largest city). But TRM is gaining wider acceptance.

On the community side, by the early 1980s, dykes were being cut by individuals and community groups – to let out water to try and reduce waterlogging, and to let in sea water and shrimp larva. People noted that breaches in the embankments allowed in sediment as had happened in the past. By the late 1980s there was community mobilization in the area, especially around Beel Dakatia where the polderization and waterlogging had caused a serious increase in poverty. In 1990, thousands of people walked 15 km to demonstrate in Khulna. There was no government response and a series of local meetings were organized. Finally on 17 August 1990, hundreds of people assembled with spades and baskets, as well as drums, pipes and other musical instruments. Working all day, they cut the embankment in four places.[38] By 1992 water levels had fallen enough to free 1,000 ha for rice cultivation.[39] However, downstream structures that had been built as part of the original project still blocked the water flows, so the cuts were not completely successful. Finally in 1994 the BWDB agreed to clear silt from downstream sluices, and water levels began to fall further.

---

37 Thomas Hofer and Bruno Messerli, *Floods in Bangladesh* (Tokyo: United Nations University Press, 2006), 100.

38 Atiur Rahman, *Beel Dakatia: The Environmental Consequences of a Development Disaster* (Dhaka: University Press, 1995), 58–60, 69.

39 Asian Development Bank (ADB) Operations Evaluation Department, *Project Performance Evaluation Report in Bangladesh* (Manila: Asian Development Bank, 2007), 40.

On the international and engineering side, the Asian Development Bank (ADB) in 1993 agreed on the $45 mn Khulna-Jessore Drainage Rehabilitation Project (KJDRP). That project was intended to clean up the mess created by USAID's Coastal Embankment Project, which ADB said had 'worsened drainage congestion and caused prolonged inundation of farmlands, household lots, and the internal communication networks. The results were declining agricultural production, fewer employment opportunities, and deteriorating salinity conditions, which collectively led to lower living standards'.[10] But the ADB's own Operations Evaluation Department in 2007 rated the KJDRP 'unsuccessful'.

Even now the international agencies and the BWDB are not fully on board, and still want to build dams and regulators. The World Bank in 2013 approved the $400 mn Coastal Embankment Improvement Project, which would try to clean up the mess caused by the ADB's KJDRP, which had tried and failed to clean up the mess caused by USAID's Coastal Embankment Project. The World Bank project has a section on 'sustainable polders', which seems contradictory. It accepts that 'polders, as they are now, are alien entities in the delta' and cannot continue. But then the project aims to 'improve agricultural production by reducing saline water intrusion in selected polders'[11] – which was the goal of the failed 1960s USAID project. The BWDB still sees itself as a big projects agency and needs large contracts to justify its existence. Some critics accuse it of choosing big contracts because they offer more opportunities for bribes and illegal fees, which are less available in TRM programmes. 'Engineers have no business unless policy makers want them to build embankments,' one proponent of new construction admitted to us, but went on to argue that 'the policy makers (and politicians) will not push for such structures unless there is popular demand for them'.

Despite the resistance of the BWDB and the World Bank, there has been a real change in attitude and an acceptance of TRM. This started in 1998 when the respected government-linked research agency, the Centre for Environmental and Geographic Information Services (CEGIS), in an ADB commissioned evaluation, backed the community option. CEGIS named it 'Tidal River Management', and found that TRM was technically feasible, socially acceptable, and environmentally sustainable compared to regulator approach promoted by the original ADB project design.[12] In the 2007 evaluation, the ADB Operations Evaluation Department said the KJDRP was unsuccessful in part due to 'continued tension between local stakeholders and the lead Executing Agency, BWDB, from the start of the Project due to diametrically opposed perspectives on the solutions to drainage congestion problems. [...] Lack of appreciation for indigenous

40 Ibid., v–vi.
41 BWDB, 'Coastal Embankment', 2. The project was approved by the World Bank in 2013 and work was expected to start in 2016, Marc S. Forni, 'Bangladesh – Coastal Embankment Improvement Project – Phase I (CEIP-I): P128276 – Implementation Status Results Report: Sequence 05' (Washington, DC: World Bank, 2015).
42 ADB, *Project Performance Evaluation*, 5, citing Environmental and Geographic Information Service (EGIS), *Environmental and Social Impact Assessment of Khulna-Jessore Drainage Rehabilitation Project* (Dhaka: Ministry of Water Resources, 1998). EGIS later became CEGIS.

knowledge systems and BWDB's resistance to adopting nonstructural solutions in favor of structural solutions were the main factors contributing to the rift between the local people and BWDB. The Project made progress only after the local people demonstrated an indigenous-knowledge-based "tidal river management" (TRM) approach, which was later found as technically feasible, economically viable, and socially acceptable'.[13]

Tidal River Management

The TRM strategy is to adopt some new methods and some of the old Mughal systems. More intensive use of land with two or three crops a year and mixed crops, fish and shrimp are part of the modern systems. Strengthening embankments to protect against cyclones is essential. But the Mughal eight-month dyke system is also incorporated – to keep out the salt water during the monsoon but allow it in during the dry season. Finally any system must maintain the relationship between polders and channels; the mud in the tide must go somewhere and if it is not dropped in a polder it will be dropped in the channel, quickly blocking the channel. If the sediment is spread over the polder, then the fast moving tide going back out to sea actually keeps the channel clear. When done correctly, TRM raises the level of the farmland and lowers the beds of the rivers and canals.

The current model, sometimes called 'rotational TRM', accepts most of the existing embankments and polders. This model takes a group of beels around a single river and allows them to flood in rotation. Up to ten beels would be sequentially flooded with muddy sea water, each for the dry season for three years, in a 30-year rotation. The sequence would need to be carefully designed, probably moving from one beel to an adjoining one and moving steadily upstream. As soon as one beel is closed another must be opened, because if there is no beel to accept the sediment, connecting rivers silt up quickly.

Decades of experimentation are still to come, as the engineers, scientists and local people all learn by doing, through trial and improvement. But TRM has been shown to work, and by starting now Bangladesh can develop systems to raise the coastal land to match sea level rise. Despite the change in mood, the engineers still want systems that require big machines. BWDB engineers argue for large-scale dredging of channels (and thus large contracts), but the Institute of Water Modelling (IWM) says this is pointless because the rapid sedimentation fills the channels within a year. The only choice is river management that maintains a high enough velocity of the flow in the channel to keep the channel clear and deep, IWM deputy director Abu Saleh Khan told us.

Dry season flooding for three years is enough to raise the level of a beel by 2 m, which should survive subsidence for two or three decades if the adjoining rivers are prevented from silting up. Flooding in just the dry season means that the beel can be used for fish, shrimp and rice for the other half of the year. Land cannot be used for three half years but the gain in level more than compensates.

43   ADB, *Project Performance Evaluation*, v–vi.

Local people saw the success of Beel Dakatia and were buoyed up by growing support of the experts. The next local action was near Beel Dakatia on 29 October 1997 when local people cut the embankments that separated Beel Bhaina from the Hari River. It was both a practical attempt to let in sediment and a political statement against the heavy engineering solutions. BWDB sued over a hundred people for damaging government property and some teachers' salaries were withheld to pay the fines. But by then the mood was changing and Beel Bhaina became a de facto experiment.[44] Sediment was deposited in the beel raising it 1 to 2 m, while 'the Hari River, which was heavily silted up before, was triple its width and its depth had increased hugely (to 10 meters deep near Sholgati Bazaar), showing again that free movement of tides will certainly keep the river channels open for drainage', according to a 2007 evaluation of the KJDRP.[15] About 600 ha were reclaimed from severe waterlogging and returned to farming; increasing employment in agriculture and fishing made a noticeable reduction in poverty.[16] The cut was finally closed on 8 December 2001. There was a quick reminder that it is essential to deal with beel and channel together, because sediment has to go somewhere. In the six months after the cut was closed, sediment accumulated quickly and the bed of the Hari River rose 6 m.

The KJDRP management accepted the success of the Beel Bhaina experiment and agreed to open Beel Kedaria, which became a tidal basin during 2002–5. The choice of this beel, 17 km upstream from Beel Bhaina, was not supported locally, and was not seen as a success. No sediment was deposited within the beel and the bed of the Hari River was not lowered. KJDRP wanted to extend the tidal flooding for another year, but the landowners refused. Beel Kedaria was too far upstream and a regulator – a dam with flaps, which allow water to flow downstream but not upstream and designed to stop salt water from entering – had been built in 1962 between beels Kedaria and Bhaina, so tidal sediment did not reach far enough upstream. Polders need to be flooded in order, downstream to upstream,[17] so Kedaria should have been the last in the sequence, not the second. The lessons of Beel Bhaina had not been learned.

The next beel to be flooded was in the right place. East Beel Khuksia is just upstream from Beel Bhaina. The first cut was made by BWDB but not where local people said it should be made. Sediment did not flow into the beel, so local people filled in the cut and made it in the place they had originally suggested. This was a success, with the ground level of the beel raised 1.5 to 2 m and the bed of the river Hari lowered 6 m – a vindication for proponents of rotating TRM. Beel Kapalia was to be next, but was stopped by the protests at Kalishakul village, mentioned at the start of this chapter.

44 de Die, *Tidal River*, 41–42.

45 ADB, *Project Performance Evaluation*, 40.

46 Alak Paul, Biswajit Nath and Md. Rana Abbas, 'Tidal River Management (TRM) and Its Implications in Disaster Management: A Geospatial Study on the Hari-Teka River Basin, Jessore, Bangladesh', *International Journal of Geomantics and Geosciences* 4 (2013), 130–31.

47 Fahad Khan Khadim et al., 'Integrated Water Resources Management (IWRM): Impacts in South West Coastal Zone of Bangladesh and Fact-Finding on Tidal River Management (TRM)', *Journal of Water Resources and Protection* 5 (2013), 959.

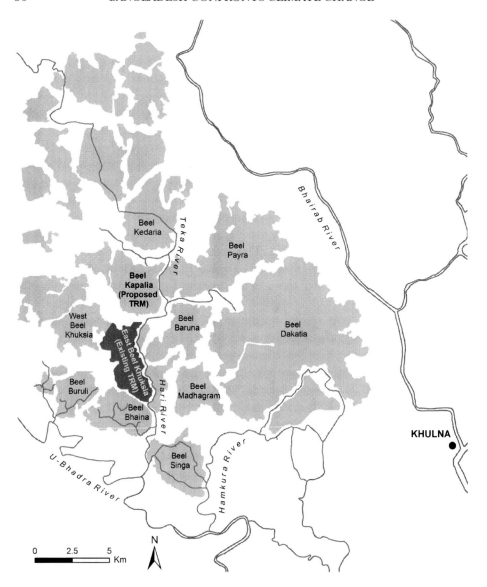

Map 4.1 Beels west of Khulna, where Tidal River Management is now being introduced.
Map: Manoj Roy

The Kalishakul Protest Shows Much Remains to Be Done

There was widespread support for tidal flooding of Beel Kapalia but the protests at Kalishakul village were triggered by worries that the mismanagement of beels Kedaria and Khuksia would be repeated.[48] With respect to both beels Kapalia and

---

48 de Die, *Tidal River*, 68.

Khuksia there had not been proper organization of local farmers; compensation had not been worked out and for some there was unemployment and increased poverty in the months of tidal flooding. Disputes between shrimp and rice farmers had not been resolved. Finally, there was continued distrust over plans by BWDB to build a new dam as part of the TRM project. In the eyes of local people, TRM has been taken over by the bureaucrats, politicians and engineers who were, yet again, ignoring local people; lip service was being given to local knowledge but local people were not being listened to.

In a thesis on TRM, Leendert de Die cites a local official as one of the organizers of the protest. The chair of Monohorpur upazila,[49] Swapana Batacharjee, said he supported TRM as the only option to drain the beels and keep the Hari River flowing, but 'we need to manage it well', which was not happening. He cited the inefficient compensation and the lack of employment for the landless during inundation.[50]

TRM has never been a simple dispute between engineers and a 'community'. It is much more complex and involves a range of interest groups at all levels, where some people benefit and some lose out. Embedded in all of this is corruption at all levels. Deregulation and contract allocation tends to favour a new elite, especially 'influential' people linked to the government of the day. The civil service is corrupted; it is widely reported that even when compensation is paid, both under TRM and for victims of cyclones, up to half the money is siphoned off by local officials. Contracts for labour-intensive works go to local 'influential' people who charge inflated prices or do shoddy work.

Some people benefit from the waterlogging caused by the dykes, for example, those who can fish on the flooded beels. This particularly helps landless people. Prof. Afsar Ali, head of Bhairab Bachao Andolon, a movement to save the Bhairab River, the river east of the Hari, said that as rivers silt up, government officials grab the new river bank land and lease it out for buildings and shrimp farms.[51] They do not support TRM because the return of the river to its former flow removes the new silted river banks.

Most significant has been the development of shrimp farming. Bangladesh has always had an inland small-scale fishing sector, with local farmers catching fish and shrimp from flooded land, ponds, rivers and canals. By 1985 the World Bank had recognized that the catch was falling due to 'deterioration of fisheries resources due to the construction of flood control works [and] lack of government commitment and support for fisheries'.[52] But with increasing export demands, shrimp culture in coastal areas and especially within polders increased, and from 1985 was promoted by the World Bank. By 2014/15, shrimp and prawn exports had reached $440 mn per year.[53] Most shrimp culture requires salt

49   The local administrative structure of Bangladesh is that the country is divided into divisions (8), then districts (64), then upazilas (sub-districts previously called thanas, 489) and at the lowest level union parishads.

50   de Die, *Tidal River*, 88.

51   'Jessore Rivers in Death Throes', *Daily Star*, Dhaka, 10 August 2014.

52   World Bank, 'Staff Appraisal Report: Bangladesh Shrimp Culture Project' (Washington: World Bank, 1985), 2.

53   Statistics Department, Bangladesh Bank. Shrimp and prawns were only 1.8 per cent of 2014/15 exports, after readymade garments ($20 bn, 83%), leather and leather products ($888 mn,

water and often competes with rice growing, leading to local conflicts. Shrimp farmers frequently cut the dykes to let in salt water at a time the rice farmers need fresh water. Conflict can be exacerbated because shrimp farmers are often better off and more politically influential. An IWM official told us that the Kalishakul protest was partly organized by 'big people who owned land they used for fish and shrimp farming' and did not want the land drained.

A similar problem involves the water management systems. A Ministry of Water Resources report noted that 'sluices are often operated by influential persons'. Large and rich land owners control when gates are opened and closed and thus control water flows. They often raise shrimp or fish or lease land to those who do, so water is regulated to benefit shrimp farmers instead of rice farmers.[54]

Compensation

To make TRM work will require intensive and complex local negotiations to share out costs and benefits. Although the land is more valuable after flooding because it is higher and not waterlogged, few people can afford to lose dry season crops for three years. Three groups need to be compensated: rice farmers, prawn farmers, and landless people and other labourers. As dry season boro rice becomes increasingly important (see Chapter 7) and less aman rice is grown during the monsoon, losses become significant. Balancing rice and shrimp is a delicate issue, and one local proposal calls for allocating less area for shrimp cultivation but with support to make them more productive,[55] which could gain the support of wealthier prawn farmers.

Compensation for poorer farmers and landless might be handled through a registration system, similar to that used for *hilsa* fishers. Hilsa is a particularly important fish, economically and culturally, and it migrates from the sea to the rivers to breed. To protect the stock, there are two closed periods during breeding when there is no fishing. Hilsa fishers are registered and during closed periods receive compensation for not fishing. The system is effective and the hilsa catch is rising. A similar registration system could be used for farm workers and other landless people from an individual beel to allow compensation during the months when the polder is subject to tidal flooding, and to give them preference for local labouring linked to the TRM.

Compensation is essential to gaining local support –making sure all affected people are compensated and limiting the amounts taken out through corruption. Land titles are often unclear, land is rented and many people are landless, so proper compensation for all affected people must be organized. This is not easy in a Bangladeshi administrative context, but there is increasing experience with cash transfer programmes, so it

   3.7%), and jute ($800 mn, 3.3%). Accessed 8 September 2016, https://www.bb.org.bd/econ-data/export/exp_rcpt_comodity.php.
54 de Die, *Tidal River*, 91 and Mainur Rashid, 'Integrated Coastal Zone Management Plan: Drainage Issues in Coastal Zone' (Dhaka: Bangladesh Ministry of Water Resources, Water Resources Planning Organization, 2005), 24.
55 *People's Plan of Action*, xx.

should be possible. Despite problems of party patronage and local corruption, existing government institutions have credibility in rural zones. Making TRM work will require strengthening the present government bodies to make them more effective and transparent.

Unfortunately, at this point the presence of donor agencies is added to the mix. International aid policy has been to weaken government structures and replace them with parallel ones, often in the form of unaccountable NGOs. One study, provocatively titled 'The Imposition of Participation', points to the failure of two decades of attempts by donors to create parallel structures such as community-based water committees that are supposed to be both non-political and self-sustaining.[56] Despite its problems, the BWDB did have a Water User Directorate with expert staff that engaged and linked with communities. But in the early 1990s the World Bank pushed the BWDB to disband the directorate. At the same time, Swedish and Dutch donors made community water management organizations (WMOs) a condition for their aid. Initially these donors promoted social mobilization aimed at giving communities some power, but they then moved to depoliticized participation more linked to communities doing the maintenance work and upkeep of water infrastructure. Also, the donors tended to work through NGOs. The study in 2012 found that more than 60 per cent of WMO executive committee members were owners of larger farms, who were less than 5 per cent of households. WMOs 'often consisted of a group of local male elites making all the decisions', and so tended to ignore uses of water important to women: for drinking and bathing and for kitchen gardens and livestock. The study also found that WMOs were ineffective in representing local knowledge – in several places infrastructure was built in the wrong place because elite demands overruled broader-based community views.

In fact, WMOs have 'proven unsuccessful in ensuring equitable water management', entrenching the power of the elites and marginalizing the voices of the poor and women, but without properly maintaining the water infrastructure. These community committees were largely dependent on donor funding, and most collapsed within two years of the end of aid funding. Interviews found in contrast to the lack of confidence in externally initiated WMOs, 'respondents from various categories voiced that they perceived the UP [the lowest level of government, the Union Parishad] as locally accessible, accountable and working for the local community.'

The UP is already the main point of contact during disasters, is important for rural labour schemes that are already doing some maintenance on water infrastructure, and is often used to settle water disputes. Rather than creating parallel ineffective systems, it would make more sense to strengthen existing local government. It is also important to

56 Camelia Dewan, Marie-Charlotte Buisson and Aditi Mukherji, 'The Imposition of Participation? The Case of Participatory Water Management in Coastal Bangladesh', *Water Alternatives* 7 (2014): 342–66 and Camelia Dewan, Aditi Mukherji and Marie-Charlotte Buisson, 'Evolution of Water Management in Coastal Bangladesh: From Temporary Earthen Embankments to Depoliticized Community-Managed Polders', *Water International* 40 (2015): 401–416.

recognize that water system maintenance cannot be locally funded, and central government must set up a permanent maintenance fund channelled through the UP.

Cyclones and Storm Surges

Another big threat overhangs the entire region: cyclones and their storm surges, which can be devastating. Of the dykes built as part of the various programmes, 957 km are sea dykes, which should also protect against storms.[57] But the BWDB admitted in 2015 that USAID's initial Coastal Embankment Project 'was designed originally to protect against the highest tides, without much attention to storm surges. Recent cyclones brought substantial damage to the embankments and further threatened the integrity of the coastal polders'.[58] Cyclone Aila in May 2009 breached the embankments in a number of places. The World Bank Coastal Embankment Improvement Project, approved in 2013, includes making the sea-facing dykes higher, wider and stronger, and reinforced where necessary. Trees will be planted to increase protection – mangroves and other salt-tolerant varieties on the shore between dyke and sea, and on the inside deep-rooted traditional trees that can withstand cyclonic wind. These will sometimes replace shallow rooted but more productive trees planted in recent years – often planted with the encouragement of development workers who ignored indigenous technical knowledge.

There is a broad acceptance of the need for strong dykes to protect the coastal areas from storm surges, which are expected to become much worse with climate change. But such dykes must be built in a way to not block the normal entry of the tides and sediments as part of TRM. This remains a crucial issue.

Conclusion: Raising the Land to Meet Rising Seas

The coastal south-west of Bangladesh, with a quarter of the country's population, may be coming to the end of misguided attempts by the engineers and donors to tame nature. As the ADB evaluation pointed out, 'strong public opposition to the proposed engineering solutions and protests against BWDB project work compelled BWDB to recognize the merits associated with the people-initiated rotational TRM to alleviate drainage congestion in the project area'.[59] The argument is not won yet. In 2007 the ADB evaluation continued that 'BWDB still believes that a large structural approach is required for solving drainage congestion and waterlogging, a position vehemently opposed by local civil society groups and NGOs as well as CEGIS.'[60] A decade later, this is still true. But the

---

57 Rashid, 'Integrated Coastal Zone Management', 10.
58 Bangladesh Water Development Board (BWDB), 'Bangladesh Coastal Embankment Improvement Project Phase 1: Short TOR on Component C3: Long Term Monitoring, Research and Analysis […]' (Dhaka: BWDB, 2015), 2. Accessed 8 September 2016, http://www.bwdb.gov.bd/tender_doc/4415.pdf.
59 ADB, *Project Performance Evaluation*, 8.
60 Ibid., 19.

mood has changed. From policy makers to local people, TRM has become the accepted model. In 2013 a *People's Plan of Action* was published jointly by the two main local NGOs, Uttaran and the Paani Committee, and the two most important national research institutes in this area, CEGIS and the IWM. 'We believe that indigenous knowledge along with academic knowledge is important for any sustainable plan,' explains Uttaran director Shahidul Islam.[61] Of course, it is not a total return to eighteenth-century eight-month embankments. New ways of using that historic knowledge are allowing the land to produce more food and support more people in the face of increasing environmental challenges.

In one of the nations most vulnerable to climate change, the southwest is the most vulnerable, facing two big threats – sea level rise and stronger and more destructive cyclones. Bigger cyclones will be met by higher and stronger dykes, and by better shelter and warning systems (see Chapter 5). Local people have shown that they can raise the level of the land, by a metre or more, which provides a way to meet the challenge of rising sea levels. In this most vulnerable place, people are showing how they can take action to keep their heads above water.

The *New York Times* repeated the received wisdom when an article on Polder 32 (see Box 4.2) was headlined 'As Seas Rise, Millions Cling to Borrowed Time and Dying Land.'[62] It said, 'Climate scientists predict that this area will be inundated as sea levels rise and storm surges increase'; indeed, climate scientists did predict this – if nothing is done to restrict global warming. But there is a choice. If global warming is curbed at 1.5°C or even 2°C, and if dykes are strengthened and a well organized tidal river management strategy is used to raise the land, then in 50 and 100 years some of the grandchildren and great-grandchildren will still be farming this land.

People in Bangladesh, from the government to the river scientists to the local communities, are doing their part. They are raising the land to match sea level rise. But in each beel, the people will need compensation for three years – to respond to damage caused by donor projects and sea level rise caused by the industrialized countries. Should Bangladesh have to pay these costs? If 'donors' pay the costs, will they continue to impose their misguided ideas of dykes and self-funding water committees, or will they let Bangladeshis use their own knowledge and ability to experiment? And, most importantly, will the industrialized and newly industrializing countries curb their emissions of greenhouse gases, or will they let the sea level rise faster than Bangladeshis can raise the land?

The millions of Bangladeshis in the coastal region have shown they can keep their heads above water – but not on their own.

---

61 *People's Plan of Action*, ii.
62 Gardiner Harris, 'As Seas Rise, Millions Cling to Borrowed Time and Dying Land', *New York Times*, 29 March 2014. A1.

Box 4.2 Sediment is 'the silver lining'

'The silver lining for Bangladesh and the delta system remains the one billion tons of river sediment that may be effectively dispersed onto the landscape,' concludes a detailed study of Polder 32 and the adjoining Sundarbans published in 2015 in *Nature Climate Change*.[63] It underlines the damaging impact of dykes and polders, but reinforces the view that not only can the land be raised to match sea level rise, but the Sundarbans, too, will rise naturally and not be flooded.

Polder 32 is an 8000 ha island with 33,000 people located 40 km south of Khulna and on the northern edge of the Sundarbans. An estimated 30–40 per cent is permanently waterlogged.[64] Cyclone Aila in 2009 breached the dykes in five places and the flooding displaced most of the population.

Aila was only a category 1 cyclone, with maximum winds of 92 km/h, compared to the category 5 cyclone Sidr two years earlier with 223 km/h winds.[65] The storm surge height was only 0.5 m above spring high tide. So why were the dykes breached? Researchers found that four of the five failures occurred at the mouths of former tidal channels that had been blocked by the dykes and that all breach sites had experienced riverbank erosion – as if the river was trying to reopen the channels. And, indeed, once the dyke was breached, the water scoured out the channel.

The dykes were not repaired for three years. The study found that between the time the dykes were built and the time of Aila, the land level inside the polder had fallen by 1.4 m – mainly due to the natural compacting of the delta soil not being replaced by new sediment. But in the two years that the breaches were open, the tides carried in sediment that raised the polder by 40 to 70 cm – exactly as would be predicted under TRM.

Land can be raised to match sea level rise.

---

63 Auerbach, 'Flood Risk': 153–57 and 492–93.

64 Bangladesh Ministry of Water Resources Water Development Board, 'Coastal Embankment Improvement Project Phase-I: Environmental Impact Assessment of Polder 32' (Washington: World Bank, 2013), xx, xxi, 2. Accessed 8 September 2016, http://www-wds.world-bank.org/external/default/WDSContentServer/WDSP/IB/2013/08/28/000356161_20130828144409/Rendered/PDF/E41410v10CEIP0000PUBLIC00Box879813B.pdf.

65 Department of Disaster Management, *Disaster Report 2013* (Dhaka: Department of Disaster Management, Ministry of Disaster Management and Relief, 2014), 22.

Chapter Five

SAVING LIVES WITH CYCLONE SHELTERS

Up to 500,000 people died in the 12 November 1970 Bhola cyclone, making it the most deadly cyclone in recorded history.[1] The lack of warning and the delayed relief operations brought huge criticism of the Pakistan government and was partly responsible for the opposition Awami League's landslide election victory, and when the AL was not allowed to take office, for the Bangladesh Independence War.

Thus saving lives in cyclones became a political priority. Since then there have been two super cyclones, Gorky on 29 April 1991, which killed 138,000 people, and Sidr on 15 November 2007, which killed 3,363 people.[2] Sidr killed only 1 per cent of those who died in Bhola – still too many, but a dramatic reduction. This success is not well known internationally, because it is the result of a largely Bangladeshi-designed and organized programme; there has been significant aid funding, but less publicity because aid agencies cannot claim it as their success.

The falling death toll is the result of a combination of cyclone shelters, sea-facing dykes and early warning systems. A first phase started after independence. A second phase started after Gorky with the Multipurpose Cyclone Shelter Programme – building many more shelters and using them as schools or offices to ensure maintenance. After Sidr a third phase has involved more shelters combined with the Second Primary Education Programme, which called for 924 dual-purpose cyclone shelter primary schools. This third phase has also involved looking at why some people do not go to shelters and included improved facilities for women and improved local warning systems. This is an evolving system, driven by Bangladeshi researchers, engineers and government, and it will have to continue to evolve to cope with climate change.

Perhaps surprisingly there is no agreement on how many cyclones have hit Bangladesh. The government's Department of Disaster Management (DDM) reports that in the 54 years 1960–2013, 45 cyclones were recorded, of which 14 were rated 'severe' and 8 of those were considered to be 'catastrophic'.[3] Hugh Brammer reports for the same

---

1   https://web.archive.org/web/20120105060230/http://www.armageddononline.org/the-worst-disasters.html last updated 15 June 2007, accessed 26 April 2016. It was probably the world's seventh deadliest recorded natural disaster. Cyclones were not named at that time, and this was given the name of the district where it made landfall.

2   Department of Disaster Management, *Disaster Report 2013* (Dhaka: Department of Disaster Management, Ministry of Disaster Management and Relief, 2014), 22.

3   Ibid., 22.

Table 5.1 Cyclone definitions and cyclones that made landfall in Bangladesh, 1998–2015

Storm category	Maximum wind speeds	Bangladesh cyclones 1998–2015[a]
Deep depression	51–61 km/h	
Cyclone	62–88 km/h	2000, 2007 (Akash), 2008 (Rashmi), 2009 (Bijli), 2015 (Komen)
Severe cyclone	89–117 km/h	1998, 2009 (Aila), 2013 (Mahasen)
Very severe cyclone	118–220 km/h	(none)
Super cyclone	222 km/h or more	2007 (Sidr)

[a] The Regional Specialized Meteorological Centre for Tropical Cyclones over North Indian Ocean of the India Meteorological Department has published good annual reports since 2005. Accessed 8 September 2016, http://www.rsmcnewdelhi.imd.gov.in/index.php?option=com_content&view=article &id=30&Itemid=176&lang=en and Wikipedia has a good series of annual pages on the North Indian Ocean cyclone season going back to 1988.

period that there were 38 cyclones of which 26 were rated 'severe'.[1] So, on average, Bangladesh is hit by one cyclone a year and by a severe cyclone once in two to four years. The frequency of cyclones is not increasing and may actually be decreasing, and this corresponds to IPCC projections (see Chapter 2). But the forecast is that cyclones will be more severe. In the 18 years 1998–2015, there were only nine cyclones, but four of those were severe (see Box 5.1). Cyclones draw their energy from the sea, and global warming means increased sea temperatures. This means stronger winds. But the big killer in cyclones is the storm surge as water is pushed up the funnel shaped Bay of Bengal into shallow coastal areas and then up rivers and canals, causing rapid flooding. In the worst cyclones, the surges can be up to 7 m high. Super cyclone Gorky in 1991 had a wind speed of 225 km/h, and modelling by Anwar Ali concluded that a 2°C temperature rise would raise the wind speed from 225 km/h to 248 km/h, and would also raise the storm surge by 1.6 m.[5]

DDM considers 37 upazilas in 12 districts to be vulnerable to cyclone, and in 2011 they had a population of 10.5 million.[6] The World Bank estimates that by 2050, climate change will raise the vulnerable population to 17 million people.[7]

4   Hugh Brammer, *Bangladesh: Landscapes, Soil Fertility and Climate Change* (Dhaka: University Press, 2016), 144, table 8.15. Other studies give other numbers, but all in the same range.
5   Anwar Ali, 'Climate Change Impacts and Adaptation Assessment in Bangladesh', *Climate Research* 12 (1999): 109–116.
6   Department of Disaster Management, *Emergency Preparedness Plan for Cyclone* (Dhaka: Department of Disaster Management, Ministry of Disaster Management and Relief, 2013), 11.
7   Maria Sarraf, Susmita Dasgupta and Norma Adams, '*The Cost of Adapting to Extreme Weather Events in a Changing Climate*', Bangladesh Development Series paper 28 (Dhaka: The World Bank, 2011), xvi.

Box 5.1 Local radio saves lives

Lokobetar is the community radio station in Barguna district on the southern coast east of the Sundarbans. One of 16 new community radio stations, Lokobetar was founded in 2011.[8] It can be heard within a radius of 17 km and covers half the district.

On Saturday 11 May 2013 a storm in the Bay of Bengal had become strong enough to be a cyclone and was named Mahasen[9]; cautionary signal 3 was issued by the Storm Warning Centre (SWC) of the Bangladesh Meteorological Department (BMD). The Cyclone Preparedness Programme (CPP) control rooms in upazilas were opened and volunteers went out to warn the community.

The next day Lokobetar broadcast a first government warning that Mahasen was strengthening in the Bay of Bengal and advised people to try to save assets. Crops were estimated as 80 per cent ripened and people were encouraged to quickly harvest. People were advised to take loose branches from trees and protect vegetable gardens.[10]

With a cyclone on the way, the district commissioner and other officials were already working with representatives of the CPP and were in contact with the DDM and the BMD. Lokobetar stayed on the air continuously.

The cyclone had been moving north towards India, but on Monday morning, 13 May, when the cyclone was still 1,300 km southwest of Chittagong, it was strengthening and shifted towards the northeast, heading directly towards Barguna. The SWC raised the warning to signal 4 and Lokobetar broadcast the first warning for fishing boats to return to port, children to be kept indoors, and people to listen to radio bulletins and CPP volunteers.[11] By now there were 5,580 CPP volunteers, plus Radio Lokebetar volunteers, spreading the message. Boats and rickshaws were mobilized to transport those who could not reach shelters on their own. It was expected that the cyclone would be 'severe' but not 'very severe' with maximum winds of 88 km/h. Most volunteers used their personal mobile phones to keep in touch with the central office and other volunteers. Increasingly residents were using radio and TV to keep up with warnings.

As the cyclone strengthened, the warning was raised to 'great danger signal 7' at 09.00 on Wednesday, 15 May, which meant everyone should take shelter. A storm surge of water 3 m above normal tide was predicted. Cyclone shelters were opened and it was announced that public buildings and some private buildings, such as hotels, were to be treated as evacuation centres. Barguna had been directly hit by cyclone

8 http://www.bnnrc.net/network/communityradioinbangladesh/lokobetar/-radio-lokobetar and http://www.bnnrc.net/network/communityradioinbangladesh/lokobetar, accessed 2 March 2016.

9 Later renamed Viyaru in India and Sri Lanka.

10 Most of this box is based on *Disaster Report 2013* (Dhaka: Department of Disaster Management, Ministry of Disaster Management and Relief, 2014), 30–31, and Amirul Islam Khan, 'Assessment Stakeholders' Role in Preparation for and Facing the Tropical Storm Mahasen' (Dhaka: Comprehensive Disaster Management Programme, Ministry of Disaster Management & Relief, 2013).

11 Disaster Management Bureau, *Standing Orders on Disaster* (Dhaka: Disaster Management and Relief Division, Ministry of Food and Disaster Management, 2010), 192–95.

Sidr in 2007 so this time people did go to the 324 cyclone shelters in the district. Indeed, DDM admits there was not enough space, and shelters became overcrowded.

Cyclone Mahasen passed close to the south of Barguna on Thursday morning 16 May at 06.00, with wind, storm surge and very heavy rain of over 200 mm. In Barguna, 7 people died, 6,900 houses were destroyed and 62,000 were damaged, and there was substantial loss of crops. But the death toll was limited by the warnings and large numbers of people taking shelter. The secretary of the local fishing association, Golam Mostafa Chowdhury, said 'the continuous broadcasting of weather news by Radio Lokobetar saved more than 2,000 trollers and trawlers.'[12]

Table 5.2 Cyclone warning signals

Signal Number	SWC label	SWC warning time	Signal label	Message and action
1–3	Alert		Cautionary	Storm in Bay of Bengal
4	Warning	24 hours	Local warning	Deep depression – smaller boats return to port; community listen to forecasts; government officials and volunteers mobilised
5	Disaster	18 hours	Danger	Cyclone – flood and wind damage predicted; all boats return to port; women, children, elderly and disabled prepare to take shelter
6–7		10 hours	Danger	Severe cyclone – evacuate to safe buildings or cyclone shelters
8–10			Great danger	Very severe cyclone – severe damage predicted; evacuation of all people

Source: SWC = Storm Warning Centre of Bangladesh Meteorological Department.
Disaster Management Bureau, *Standing Orders on Disaster* (Dhaka: Disaster Management and Relief Division, Ministry of Food and Disaster Management, 2010), 192–95.

---

12 Trollers are fishing boats which pull long lines behind them, while trawlers pull nets.

Shelters

Records of devastating cyclones go back to 1582 and 1699, and in the colonial era more than 100,000 people were killed in each of three cyclones, in 1876, 1897 and 1911. In the Pakistan era the first steps were taken. As noted in Chapter 4, the Krug mission proposals for dyke and polder systems proved to be misguided, but his proposals for sea-facing dykes in the coastal embankment project began to be implemented in the 1960s and provided important protection. These embankments do not keep out the water, which can go around and behind them, but they obstruct the penetration of the surge water and reduce its energy.

The first purpose-built cyclone shelters were constructed in the 1960s.[13] But it was only after the disaster of super cyclone Bhola in 1970 and the subsequent liberation war that independent Bangladesh moved forward seriously with cyclone protection. This involved continuing with dykes, developing cyclone shelters, creating proper early warning systems and more recently improving the security of people's own homes. The World Bank estimates that the government has invested more than $10 bn in cyclone and flood protection but that it needs to invest at least an additional $2.5 bn. Climate change will add another $2.4 bn, including $1.2 bn for more shelters and $0.9 bn for 33 sea-facing polders where dykes will be overtopped by storm surges caused by the stronger cyclones and must be raised and strengthened.[14]

The CPP was launched in 1972 and is a collaboration between the government and the Bangladesh Red Crescent Society (BDRCS). After each of the super cyclones, Gorky in 1991 and Sidr in 2007, there were evaluations of cyclone protection and both led to changes in systems and structures. In 2012 a separate Ministry of Disaster Management and Relief was created, and is staffed by some of the country's most dedicated and skilled civil servants.

Cyclone shelters are large concrete structures built on reinforced concrete pillars and the entire ground floor is left clear to allow wind and water from a storm surge to pass underneath. There are typically one or two upper stories, and shelters can hold up to 2,000 people. There is no single standard design, but many are arrow-shaped with two wings and the arrow pointing south into the wind. Design standards assume 4–5 people per m², which is the density of people standing on a bus or in a lift, and is thus very crowded.[15] By 1992, 512 cyclone shelters had been constructed, some by the BDRCS.[16] After Gorky the government's planning commission introduced the Multipurpose

13 Atiq Rahman and Rafiq Islam, 'Shelters and Schools – Adapting to Cyclonic Storm Surges: Bangladesh' in *Climate of Coastal Cooperation*, ed Robbert Misdorp (Leiden: Coastal & Marine Union – EUCC, 2011), 172 and Zheng Jia, 'Cyclone Shelters and Cyclone Resilient Design in Coastal Areas of Bangladesh', Master in City Planning thesis, Massachusetts Institute of Technology, 2012, 41.

14 World Bank, *The Cost of Adapting to Extreme Weather Events in a Changing Climate* (Dhaka: World Bank, 2011), Bangladesh Development Series Pay No. 28, xiii, xvii, 30.

15 Zheng Jia, 'Cyclone Shelters', 43; and Zheng Li and David A. Hensher, *Crowding in Public Transport: A Review of Objective and Subjective Measures* (Sydney, Australia: Institute of Transport and Logistic Studies, University of Sydney, 2012).

16 Department of Disaster Management, *Disaster Report 2013*, 30.

Cyclone shelter under construction in Dacope, adjoining a settlement of new houses for people displaced by a previous cyclone. The ground floor is open to allow wind and water to pass through. The first floor will be the shelter, and also be the local school classrooms. Photo: Joseph Hanlon

Cyclone Shelter Programme to develop an effective cyclone shelter protective network, drawing on experts from Bangladesh University of Engineering and Technology and Bangladesh Institute of Development Studies. By the time of Sidr there were 2,400 shelters.[17] But even that was not enough. For example, in Barguna district (see Box 5.3) the capacity of the shelters at 5 people per m² is 206,065, but 255,300 people crammed into the shelters during cyclone Sidr.[18]

The Second Primary Education Programme in 2007 linked new schools to shelters in coastal areas, and planned 924 joint schools and shelters. There was an emphasis on improved facilities, with larger classrooms, improved toilet facilities and safe drinking water. By 2014 the number of shelters had increased to 3,777 and it was predicted that at least 700 more shelters would be built by 2017. But this will still be only half the number of shelters needed. Moreover, the World Bank estimates that climate change will require that Bangladesh build an additional 5,700 shelters by 2050 at a cost of $1.2 bn.[19]

Shelters clearly work. Bimal Kanti Paul wrote that a study after Sidr in 2007 'reported no fatalities among people who opted to move to [...] a public cyclone shelter. All deaths seem to have occurred among people who did not comply with evacuation orders, were turned back from shelters (many of those facilities were already full and even over-crowded) or who took shelter in the structurally more strong homes of neighbours, friends and relatives'. Indeed, 17 per cent of those interviewed said they were turned

17 Bimal Kanti Paul, 'Why Relatively Fewer People Died? The Case of Bangladesh's Cyclone Sidr', *Natural Hazards* 50 (2009): 297.

18 Zheng Jia, 'Cyclone Shelters', 65, citing Center for Environmental and Geographic Information Services, 'Report on Cyclone Shelter Information for Management of Tsunami and Cyclone Preparedness' (Dhaka: Comprehensive Disaster Management Programme, Ministry of Food and Disaster Management, 2009).

19 World Bank, 'The Cost of Adapting', xvii.

away from overcrowded shelters.[20] Up to 40 per cent of coastal residents evacuated before Sidr, with half going to shelters and the other half taking shelter in neighbours' stronger houses, public buildings such as school and mosques, or up on embankments and other higher land.[21]

Shelter design and management has changed over the years, reflecting research and reports from users. The first problem to be noted is that shelters are only used once a year or less, and that there was no maintenance budget or system. After Gorky that led to the Multipurpose Cyclone Shelter Programme with the idea that the buildings could be used as schools or offices, which would ensure they were maintained. A report in 2009 found that 82 per cent of shelters were used as education centres, 8 per cent as offices, 1 per cent as community centres, and 1 per cent as health centres. Only 6 per cent did not have any non-emergency use.[22]

The 2012 design standards[23] introduced a new generation of shelters, with shelters expected to withstand wind of 260 km/h and 6 m high tidal storm surges; the design standard is three people per m², about 170 people in a school classroom. The new guidelines take into account prior experience and especially that many people did not take shelter even if they heard the warnings. Partly there were simply not enough shelter spaces. But it also became clear that women often failed to take shelter. In part this was because shelters did not provide separate space and a toilet for women; there was also lack of water and electricity. Shelters are being modified to have these facilities and it was added to the design of new shelters, but the 2009 survey showed that most of the shelters still had not done this. The 2012 design standard requires a separate room and toilet facilities for women and separate toilet facilities for pregnant women. New shelters must have solar panels and rainwater harvesting systems (because local ground water may be saline). The other issue is that women are responsible for the household, including animals, which they did not want to leave unattended. Cyclone Sidr killed 100,000 cattle, buffalo, sheep and goats.[24] This led to another change, first introduced by the Bangladesh Red Crescent and now part of the design standard – to build livestock pens called *killas* on earthen platforms next to the shelter or to use the space under the shelter to hold 300–500 cattle. Another issue was that shelters failed to provide access facilities for the disabled;[25] thus the new guidelines require a ramp and

20 Paul, 'Why Relatively Fewer People Died?' 290, 298, 302.

21 Bimal Kanti Paul et al., 'Cyclone Evacuation in Bangladesh: Tropical Cyclones Gorky (1991) vs. Sidr (2007)', *Environmental Hazards* 9 (2010): 89–101.

22 Center for Environmental and Geographic Information Services (CEGIS), *Cyclone Shelter Information for Management of Tsunami and Cyclone Preparedness* (Dhaka: Comprehensive Disaster Management Programme Ministry of Disaster Management and Relief, 2009), 21.

23 Ministry of Disaster Management and Relief, 'Cyclone Shelter Construction, Maintenance and Management Guideline 2011' (Dhaka: Ministry of Disaster Management and Relief, 2012). Accessed 14 April 2016, http://modmr.portal.gov.bd/sites/default/files/files/modmr.portal. gov.bd/publications/73ed76b0_385a_438c_bf25_b9bc2acca079/Cyclone%20Shelter%20 Construction%20Maintenance%20and%20Management%20Policy%202011%20(1).pdf.

24 'Cyclone Sidr Causes Tk130cr Livestock Loss', *Star*, Dhaka, 7 January 2008.

25 CEGIS, *Cyclone Shelter Information*, 22.

that 'a reasonably sized room should be kept reserved for the disabled'. Some shelters now have storage facilities for the valuable goods of people who go there during a cyclone. Sometimes the shelters are themselves built on platforms. Some people do not move to shelters until the surge water is in their courtyards, so the final design change has been to ensure that approach paths are all paved and on embankments above the expected water level.

Cyclone Warnings

The government of Bangladesh has developed an impressive forecasting and warning system. The Storm Warning Centre (SWC) of the Bangladesh Meteorological Department (BMD) issues storm warnings including observed and predicted path of the cyclone and, as it gets closer, predictions of wind speed, strong wind zone and storm surge height. This is based on satellite monitoring, radar, reports from the Regional Specialized Meteorological Centre for Tropical Cyclones Over North Indian Ocean (in India) and, increasingly, computer models. However, as BMD says, 'conventional methods are used for analyses and subjective forecast is made. So, forecasting depends on personal skills'.[26] Some of the early development of forecasting was done in the 1980s with significant help from the World Meteorological Organization, but Bangladesh has moved on to become one of the most advanced countries in cyclone tracking and prediction.

In the giant government complex in central Dhaka, the Bangladesh Secretariat, is a room no bigger than a school classroom with some computers, which is the Emergency Operation Centre. When a cyclone is threatened, the action moves there. When a signal 4 is issued by BMD, all ministry permanent secretaries move to this room, and the cyclone becomes their main task. At signal 7, their seats are taken by the ministers themselves – and no one leaves until the cyclone is over. It is a system in force since 2012 and is linked to the decision in 2012 to establish a Ministry of Disaster Management and Relief to strengthen coordination. Cyclones Komen (2015) and Mahasen (2013) reached signal 7 and demanded the presence of the ministers.

But forecasting is only useful if people can be informed. This is the role of the 50,000 volunteers (of whom one-third are women)[27] of the CPP, run jointly by the government and the BDRCS in the 37 cyclone vulnerable upazilas. The normal system has been that bulletins are transmitted to the 6 zonal offices and the 30 upazila level offices (sub-district) over HF radio, and the upazila offices pass information to unions and lower levels through VHF radios. There is now mobile telephone coverage of the entire country and

---

26 Sayeed Ahmed Choudhury (Meteorologist, Bangladesh Meteorological Department), 'Country Report of Bangladesh On Effective tropical cyclone warning in Bangladesh', paper presented at JMA/WMO workshop on effective tropical cyclone warning in southeast Asia, Tokyo, Japan, 11–14 March 2014.
27 In 2014 there were 49,365 volunteers of whom 32,310 were men and 16,455 were women, http://cpp.gov.bd/ and http://www.bdrcs.org/programs-and-projects/cyclone-preparedness-program, accessed 1 March 2016.

most volunteers have mobile phones and are using them. SMS (text) warnings were first used for cyclone Aila in 2009. E-mail and a website are also used.

Volunteers are given a kit including a transistor radio (for their own information), a megaphone or loud hailer and a hand siren (to use for signal 7, mandatory evacuation). Mosques also broadcast warnings. A study by Bimal Kanti Paul found that 86 per cent of all households were aware of the cyclone warning for Sidr.[28]

However, information sources are changing. More people gain information from TV or radio, including community radio (see Box 5.3). In 2011 mobile telephone companies introduced an interactive voice response (IVR) line. Dialling 10941 gives an up-to-date warning message. On 15 May 2013, 29,369 people phoned the line.[29] But surveys show that despite people being able to obtain information on their own, volunteers remain important for mobilizing people to take shelter and often helping them to reach the shelters.

Risk Reduction

After Sidr and as part of a global change in thinking, the government moved more towards a strategy of risk reduction and towards more involvement of communities with the recognition that government cannot do everything on its own – which also recognizes the centrality of the CPP volunteers. In cyclone zones this meant recognizing that not everyone goes to shelters. Cyclone shelters are designed to serve everyone within 1.5 km, but in very high wind and heavy rain, even that distance can be difficult to walk, and many people wait until the last minute because shelters are crowded and uncomfortable. Also people are reluctant to leave their homes for a variety of reasons, including fear of theft of their limited household goods. Nevertheless, it is estimated that only 5 per cent of houses in poor coastal areas can withstand serious storm surges.[30] This has led to increased efforts to develop small shelters and more robust houses. Because the storm surge is the main killer, houses need to be built on mounds or plinths above the expected water level; this extends a practice already widely used, but often means raising the plinth level, which is expensive. Toilets and water pumps should also be on raised ground so that they do not flood.

Various experiments and pilot projects are now being carried out by NGOs and by the BRAC University Architecture Department. A cyclone resistant pucka house made of brick and concrete is too expensive for most rural people. One alternative model is to build a concrete and brick core house, of perhaps one room. We visited a donor-funded project in Dacope upazila near the Sundarbans where plinths had been built and which had a core house plus toilet and pump above water level. People had expanded the houses in varied ways using traditional bamboo and thatch construction.

---

28 Paul, 'Why Relatively Fewer People Died?' 297.
29 Amirul Islam Khan, 'Assessment Stakeholders' Role in Preparation for and Facing the Tropical Storm Mahasen' (Dhaka: Comprehensive Disaster Management Programme, Ministry of Disaster Management & Relief, 2013), 5.
30 Paul, 'Why Relatively Fewer People Died?' 291.

Experimental cyclone resistant core house in Dacope. The small concrete core house is on a platform to raise it above flood level. On the far left is a tank to collect rainwater, and out of sight is a latrine also on a platform above flood level. The family has then built additional rooms (right) with conventional construction, but raised only slightly above ground level. Note that the roof is covered with pumpkin vines, but there is still space for a solar panel. Photo: Joseph Hanlon

Many had built the extra rooms at ground level, off the plinth, but linked to the core house. Others had squeezed additional rooms onto the plinth. One had built a wide veranda around the core house, linked to other rooms, with the traditional pumpkins growing on the roof.

A core house is only a form of risk reduction – in the event of a cyclone the family shelters in the strong core but may lose part or all of the additional traditional construction. The core house appears to cost only $6,000 – too expensive for most local people, but not too expensive for government or donors as part of a cyclone resilience project. In South Africa the government has built 3 million core houses measuring [31] to 36 m² in two decades at a cost of about $8,000 each, for low-income people, so it is possible in Bangladesh too.

Research into design of traditional structures and work with local builders has also led to some recommendations. In a cyclone, roofs lift off houses for the same reason that

31 Known as 'RDP houses' after South Africa's post-apartheid Reconstruction and Development Programme.

airplanes fly – the wind going over the top creates a suction that lifts the plane or the roof. This can be reduced if the main roofs are built with a lower pitch, between 25 and 35 degrees. It is important to make sure the foundations, roofs and walls are firmly fixed together, perhaps using galvanized wire instead of jute ropes. Perhaps most important is to recognize that a triangle is a much more stable form than a rectangle, which can easily bend into a parallelogram.

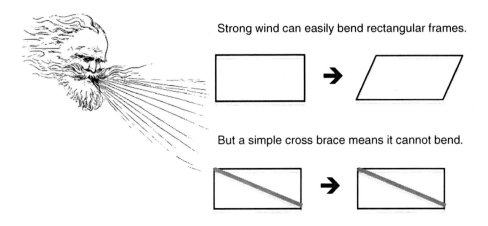

Strong wind can easily bend rectangular frames.

But a simple cross brace means it cannot bend.

The normal frame of a house is made up of rectangles of bamboo poles, but adding this simple brace to each rectangle makes a large difference. Better anchoring of bamboo poles into the ground with brackets or concrete also improves strength. All of these adaptations cost money, but relatively little, and as cyclones become stronger they are worth the investment.

There are many pilot and experimental projects constructed by NGOs with no coordination and little shared information. The House Building Research Institute (HBRI) of Bangladesh is trying to do some coordination.

Another important area of adaptation relates to trees. People have known for centuries about the importance of trees as windbreaks. Traditionally, rural Bangladeshis leave the south and east sides of the houses open to allow sunlight and breeze to enter, but plant higher trees on the west and north as a windbreak. Part of the new move for adaptation and risk reduction is to plant trees again, further protecting houses. Trees around embankments, killas and cyclone shelters reduce the impact of high winds. They also serve as an impediment to waves and storm surges and help to stabilize the ground and embankments. Choice of trees is important; there had been a move to plant trees that are productive but not native to the area. Some of these trees tend to have shallower roots and cannot withstand cyclones as well as the traditional trees, which are now favoured to resist winds.

The other area for tree planting is on the sea-facing side of dykes and embankments. The Sundarbans showed the importance of mangroves in reducing the force of the wind and surge of cyclone Sidr. Reforestation where many trees have been cut in the

Sundarbans is an important new policy, as is planting mangroves and other trees on the sea-side of embankments. This is another example that granddad knew best.

Planting deeper-rooted trees and building stronger houses are not as dramatic as large cyclone shelters, but they are sensible responses to a slow onset disaster. As climate change makes cyclones stronger and wind speeds and surge heights greater, these incremental changes can reduce the damage caused by the strengthening of cyclones.

Typically ignored by NGOs, international agencies, donors and the government is depreciation and maintenance. Cyclones make landfall on only part of the coast, which means cyclone shelters and stronger houses are only used once or twice in a decade. But they must be checked and food and water stockpiled before each cyclone season. Essential maintenance must be done to these buildings every year. Some of this can be done by the schools or other occupants of shelters. But major maintenance, including structural repairs, will be needed every 7–10 years. Who organizes and pays for this? NGOs and other aid programmes that built the shelters or strong core houses will have moved on to other programmes long before major repair works are done. Bangladesh does not have a good record on maintenance and rehabilitation, and there have been frequent criticisms of failures to maintain dykes.

Conclusion: Home-Grown Successes

An incredible 2.6 million people used cyclone shelters during cyclone Sidr in 2007.[32] Shelters, early warning and volunteers saved tens of thousands of lives. Designed and implemented largely by national scientists, engineers, administrators and tens of thousands of volunteers, it is evidence again that Bangladeshis are not helpless victims, but instead are taking the lead in action to respond to an extremely difficult and challenging environment. These responses are steadily changing, as people learn from experience – each generation of cyclone shelters is better and the warning systems improve. It is an impressive national initiative.

Hurricane Katrina that hit New Orleans in 2005 was, in Bangladeshi terms, a 'super cyclone', like Sidr two years later. One of the poorest countries in the world coped well. The richest country did not. In New Orleans the dykes collapsed despite claims that they were built to withstand such a storm; there were no shelters and an inadequate evacuation plan. Both Sidr and Katrina were predicted well in advance. But unlike Bangladesh, which automatically brings ministers into an operations room when a super cyclone is close, a US House of Representatives committee[33] discovered that the highest level committee in the US is of government agency officials, and it was not convened for Katrina, further contributing to the disorganization. There were complaints about the slowness

---

32 Center for Environmental and Geographic Information Services (CEGIS), Cyclone Shelter Information for Management of Tsunami and Cyclone Preparedness (Dhaka: Comprehensive Disaster Management Programme Ministry of Disaster Management and Relief, 2009) 47.

33 A Failure of Initiative, Final Report of the Select Bipartisan Committee of the U.S. House of Representatives to Investigate the Preparation for and Response to Hurricane Katrina (Washington, 2006).

in repairs in Bangladesh after Sidr, but those works were completed before the ones to repair New Orleans, which had been hit two years earlier. Both countries face problems of corruption and political patronage; in 2014 Ray Nagin who was mayor of New Orleans during Katrina was sentenced to ten years in jail for taking bribes from city contractors before and after Katrina. Despite its difficulties, one of the poorest nations in the world is successfully confronting a very difficult environment.

More serious super cyclones will be one of the first signs of climate change, which means climate change will hit these 'at-risk' upazilas first. Bangladeshis know this all too well. This is one reason for the intense international campaigning and negotiating by Bangladeshis as detailed in Chapter 3. And they also know that even the present global warming, to which they have made no contribution, will increase the risk of cyclones. Increased global warming heats the Bay of Bengal, which will make the cyclones more powerful. There is a direct link between industrialized country emissions and wind speed in these cyclones; it is the cars and central heating and businesses in Europe and the United States that will directly cause strong cyclones in Bangladesh.

Here on the coast, climate change creates no new problems – but it makes the existing ones much worse and requires that existing actions be stepped up: higher and stronger embankments, more and better shelters, stronger houses and so on. It will cost more than $1 bn to build more shelters and billions of dollars more to cope with global warming to which Bangladesh has contributed little – but which will accelerate because the rich countries in Paris in December 2015 said they would keep raising the temperature.

As they turn up the heat, will the rich pay for Bangladesh to defend against the stronger cyclones? And will they let Bangladesh get on with the job, or, as with dykes and polders, will outsiders say they know better?

Chapter Six

LIVING WITH FLOODS

The three-storey municipal office building collapsed into the river in July 2013. In three years all the Chowhali upazila municipal offices, the police station, the Bangladesh Rural Development Board office, a hospital and health centres, several schools, 20 km of paved road – and all flood protection barriers – were eaten by the river.[1] The Brahmaputra-Jamuna is 12 km wide at this point and is called a 'braided river', with multiple constantly changing channels separated by alluvial islands. Chowhali upazilia is 100 km northwest of Dhaka and had been comfortably east of the river. After 2008 the river opened a small channel near the town which became a major channel and the huge increase in water flow rapidly eroded the east bank of the river and the town centre disappeared into the Jamuna river.

The Brahmaputra enters the north of Bangladesh from Assam, India. Historically it flowed east of Dhaka and into the Meghna. But in the late eighteenth century the Brahmaputra river changed course at the point where it enters Bangladesh, and shifted some of the main flow into the Jamuna river which joins the Ganges-Padma west of Dhaka. This shift is continuing – in 1830 the new and old channels seemed to be equal, but in recent years the old Brahmaputra has much less water. Therefore the Jamuna is expanding, from 5.6 km wide in 1914 to 8.5 km wide in 1973 and 12 km wide in 2010.[2] It was the widening river that devoured Chowhali upazila.

As the rivers erode the banks, some of the sediment is dropped in the channel, creating shoals that become islands known as chars. These chars are occupied by farmers, but they are under water during normal floods. Most chars last only a few years before they are washed away again, but some become permanent islands with trees and buildings. In 1992 it was estimated that nearly 2 million people lived on Brahmaputra-Jamuna chars.[3] From 1973 to 2001 the total char area in the lower Jamuna increased by two-thirds, from 424 km^2 to 710 km^2, while the total river area increased from 686 km^2 to 999 km^2.[4]

1   Md. Abdur Rahman Khan, *A Study of Chowhali Upazilla and Its Effects Due to River Erosion* (Rajshahi, Bangaldesh: Rajshahi Urban Development Directorate, 2015).

2   Maminul H. Sarker et al., 'Morpho-dynamics of the Brahmaputra–Jamuna River, Bangladesh', *Geomorphology* (2014): 45–59 and Hugh Brammer, *Can Bangladesh be Protected from Floods?* (Dhaka: University Press, 2004), 10, 50.

3   Brammer, *Can Bangladesh be Protected*, 49–52.

4   Northwest Hydraulic Consultants, Canada, 'Bangladesh: Main River Flood and Bank Erosion Risk Management Program', Asian Development Bank Technical Assistance Consultant's Report, 2013, ix.

Just 50 km to the west, on the other side of the Jamuna, the problem is just the oppo-site. Chalan Beel, one of the largest wetlands in Bangladesh, is drying out. On 14 March 2013, just four months before Chowhali was lost, there was a 220 km long human chain along the banks of the Boral River, to protest that this was once a navigable river and had now been turned into a mere trickle. Protestors blamed this on the cross dams, regulators and road embankments built in the 1980s for cutting off the flow of water, and demanded removal of the dams.

Floods and Responses

For centuries, Bengalis have been trying to understand the complex flood pattern and to respond to it. Bangladeshis distinguish between normal floods (*barsha* in Bangla), which are part of the agricultural season, and damaging floods (*bonna*). Floods are hugely vari-able from year to year (see Figure 1.1). Serious floods that cause crop damage occur every 3–5 years, while catastrophic floods causing major damage occur approximately once a decade. The 1988 flood was the worst in recorded history, only to be supplanted by the 1998 flood as the worst – both flooded more than 60 per cent of the country. Damage in the 1998 flood and the less disastrous 2004 flood were each over $2 bn, with more than 30 million people affected.[5] Water and sediment comes from three major river systems, each with its own catchment area and weather patterns.[6] To this is added Bangladesh's own heavy monsoon rainfall. Four different sources of water mean that no two floods are the same, and that levels and types of flood water are different in various parts of the country.

It is useful to divide the country according to the extent of 'normal' flooding, and according to the source of the water. Hugh Brammer in his book *Can Bangladesh be Protected from Floods?* explains that 29 per cent of the country is highland, which does not normally flood; 35 per cent is medium highland,[7] which normally floods in the monsoon to a depth of 90 cm and can be used for transplanted *aman* rice; and 21 per cent is lowland normally flooded to a depth above 90 cm and sometimes more than 300 cm[8] and for which a whole range of locally specific tall rice varieties have been developed over the centuries.

It is also useful to divide the medium highland and lowland areas according to the three different sources of flood water – silt-filled sea water, silt-filled river water, and clear rain water. As noted in Chapter 4, the southwest is not flooded by rivers, but rather

---

5   Md. Mizanur Rahman, Md. Amirul Hossain and Amartya Kumar Bhattacharya, 'Flood Management in the Flood Plain of Bangladesh' (Dhaka: Bangladesh Water Development Board). Accessed 7 April 2016, https://www.academia.edu/702051/Flood_Management_ in_the_Flood_Plain_of_Bangladesh.

6   It had been widely claimed that floods in Bangladesh were being made worse by deforestation in the Himalayas, but this is now accepted to not be true. Deforestation has been going on for centuries and most erosion is natural, not human-made. Brammer, *Can Bangladesh be Protected?* 92 and Thomas Hofer and Bruno Messerli, *Floods in Bangladesh* (Tokyo: United Nations University Press, 2006), 428.

7   Medium 'high' land is relative, and can be only 10 m above sea level.

8   Brammer, *Can Bangladesh be Protected*, 26–27.

by high tides as well as storm surges that carry silt-filled sea water. Brammer points out that 'high river levels block the drainage of run-off from rainwater falling on adjoining flood-plain land. Therefore, most of Bangladesh's floodplain areas are flooded by clear rainwater or the raised groundwater-table, not by silty river water'.[9] Thus chars and riverbanks are flooded by silt-filled river water, and the slower flowing water at the edges of channels tends to deposit sediment and raise the land level, while areas further from the rivers are flooded by rainwater that is trapped by higher water levels in the rivers and cannot escape.

In the seventeenth century during the Mughal rule of the Indian sub-continent, extensive water management systems were developed and codified in what is now Bangladesh. These attempted to make the best use of normal flooding while also controlling damaging high floods. Major James Rennell, the first British Surveyor-General of India, reports in his journals for 11 June 1764 discovering in what is now Bangladesh 'a large Dam is thrown up to keep y'' River from overflowing the Countrey in the height of the wet Season. This Dam extends more than 5 miles; it is about 12 foot high & 14 yards thick'.[10] This 8 km long, 4 m high flood control dyke was a major civil engineering structure, and was then so old that local people could not say when it had been built.

Sophisticated channel and canal systems were developed with sluices and other systems to control flow, particularly for irrigation and to drain excessive monsoon rainwater. The importance of river sediment was recognized and a system of overtopping was developed to ensure that sediment reached the fields. Reports from 1771 said that the flood control and irrigation embankments (then called bunds) that had been maintained by local rulers called *zamindars* in the Mughal period were no longer being kept in repair under British rule.[11] This led to a series of Embankment Acts to force their continued maintenance by local residents. In 1930 a British engineer, Sir William Willcocks, gave a series of lectures at Calcutta University in which he explained that the Mughals, perhaps building on even earlier systems, had created networks of long parallel canals for what he called 'overflow irrigation'.[12] River water was diverted down these canals and cuts were made in the banks to add river water to the monsoon rainwater. The silt built up the land and disrupted the breeding of malaria mosquitoes; the river water also contained fish larvae which grew in the flooded fields. When the rivers fell the cuts were closed, allowing water to remain in the fields for the rice. Later in the season the canals were used to drain surplus water from the fields. But the canals needed regular maintenance.

The canals deteriorated over time, in part due to lack of maintenance, and in part due to the construction of the railways in the late nineteenth and early twentieth century. Water transport had been the predominant means of travel and transporting goods, but

---

9 Ibid., 25.

10 T. H. D. La Touche, ed., *The Journals of Major James Rennell* (Calcutta: The Asiatic Society, 1910), 116.

11 Henry Leland Harrison, *The Bengal Embankment Manual* (Calcutta: Bengal Secretariat Press, 1875).

12 William Willcocks, *Lectures on the Ancient System of Irrigation in Bengal* (Calcutta: University of Calcutta, 1930), four lectures given in 1930.

the British colonial authorities stopped maintaining the waterways and instead built huge embankments above flood level for railways. These embankments were not built with enough drainage openings so there was waterlogging upstream of the embankments, and lack of flow meant silt was deposited. Flooding and waterlogging behind the embankments and the pits dug to build the embankments created breeding areas for mosquitoes and malaria increased, while agricultural production fell.[13] Over time the unmaintained canals changed their course and were seen as creeks and rivers, but Willcocks argued that the original artificial pattern could still be seen.

Box 6.1 Chalan Beel: A century of problems

Once one of Bangladesh's largest wetlands and an important area of farming and fishing, Chalan Beel has for more than a century been an example of what can go wrong with engineering solutions. In his report commissioned by the government of Bengal on the serious floods of 1922 and earlier, Prof. Prasanta Chandra Mahalanobis pointed to the way Chalan Beel was shrinking, due to siltation, apparently because silt-laden water from the Ganges-Padma entered the beel but could not get out. This in turn was caused by the building of railway embankments with very few openings through which water could pass.[14] The Eastern Bengal Railway main line embankment blocked many of the 47 rivers that came down from northwest Bangladesh; the branch to Sirajganj is just south of the beel and blocked the outflow to the Brahmaputra-Jamuna, which would have normally flushed out the silt. Flooding behind the embankments made it impossible to grow aman rice.[15] In 1985 sluices were built on the Boral River, which is a branch of the Ganges-Padma and which passes through Chalan Beel and then goes to the Brahmaputra-Jamuna. As Hugh Brammer commented, 'the basic project design was at fault'.[16] Activists complain that the gates had not been opened for several years, meaning no water flowed down the river, and the river banks were illegally occupied.[17] Tens of thousands of activists formed a human chain for an hour along the 220 km length of the river on 14 March 2013, to demand the removal of sluices and dams and the eviction of illegal occupiers.

---

13 Iftekhar Iqbal, 'The Railways and the Water Regime of the Eastern Bengal Delta, c1845–1943', *Internationales Asienforum*, 38 (2007): 329–52.

14 P. C. Mahalanobis, *Report on Rainfall and Floods in North Bengal 1870–1922* (Calcutta: Bengal Secretariat Book Depot, 1927), 31, 44.

15 Iftekhar Iqbal, 'The Railways', 329–52 and Iftekhar Iqbal, 'Railways and the Water Regime of the Eastern Bengal Delta, c. 1905–1943: A Reappraisal', no date. Accessed 30 March 2016, http://www.sasnet.lu.se/EASASpapers/13IftekharIqbal.pdf.

16 Brammer, *Can Bangladesh be Protected*, 228.

17 Abu Bakar Siddique, 'Remove Sluice Gates to Save River Boral: Environmentalists', *Dhaka Tribune*, 12 October 2013.

Engineering Answers

Partition and independence of Pakistan plus new global relations due to the cold war, the United Nations, and growth of foreign aid led to a Western response to damaging floods and cyclones in the 1950s and 1960s. This was an age of 'high modernism', when it was assumed that science could solve all problems and humans could control the natural environment. The common response to flooding was for structural, engineering solutions – building dykes and dams to control the water –which proved to be partly misguided. Chapter 4 looked at the problems this caused in coastal areas. Further north, 7,500 km of river embankments had been built by 1988.

Devastating floods in 1987 and 1988 drew international attention to Bangladesh. French First Lady Danielle Mitterrand flew to Bangladesh to see for herself and the French sent 30 engineers to design a permanent solution to Bangladesh's flood problem. Other donors scrambled to help, not wanting the French to take the lead. The Group of Seven (G7) industrialized countries at their meeting in Paris in July 1989 declared that 'it is a matter of great international concern that Bangladesh is periodically devastated by catastrophic floods.' There were four different plans drawn up by large donor teams, and a Flood Action Plan (FAP) was launched in London in December 1989, with the World Bank trying to sort out the competing interests.[18]

From the beginning there were sharp divisions. France and the UNDP, backed by Bangladesh's Ministry of Water Affairs, said the rivers had to be embanked and pushed for a $4–10 bn programme, which would have been one of the largest development projects ever undertaken.[19] The United States, Japan and the Ministry of Agriculture were more cautious. The Ministry of Agriculture opposed further embankment on the grounds of poor past performance and adverse environmental effects.[20] USAID said that controlling floods with river embankments was neither technically feasible nor ecologically desirable.[21] The issue was further complicated because this was during the waning days of military rule, headed by General Hussain Muhammad Ershad, which blocked public discussion of the project. Thus opposition to FAP became part of the civil society campaign for democracy.

The World Bank agreed to coordinate the London meeting. The Bank made the underlying assumption that the major rivers would have to be controlled, and largely drew on the French and UNDP studies.[22] An army of consultants was commissioned

18 Nerun N. Yakub, 'Overview', in *Rivers of Life*, ed. Kelly Haggert et al. (Dhaka: Bangladesh Institute for Advanced Studies and London: Panos, 1994).

19 David Lewis, 'The Strength of Weak Ideas? Human Security, Policy History and Climate Change in Bangladesh', in *Security and Development*, ed. John-Andrew McNeish and Jon Harald Sande Lie (Oxford: Berghahn, 2010), 113–29.

20 Yakub, 'Overview' and Brammer, *Can Bangladesh be Protected*, 168.

21 Independent Evaluation Group (IEG) of the World Bank, 'The World Bank Support for Flood Control in Bangladesh' (Washington, DC: World Bank, 1991). Accessed 5 April 2016, http://lnweb90.worldbank.org/oed/oeddoclib.nsf/DocUNIDViewForJavaSearch/C76A820FE0CBD784852567F5005D8234?opendocument.

22 IEG, 'The World Bank Support for Flood Control'.

for a five year $150 mn study phase. There were 33 different segments and hundreds of reports funded by various donor governments. Ershad initiated his own Dhaka embankment project with government funds after the 1988 flood. But public protests were growing, and with the end of the cold war the United States reduced its support for dictators. Ershad was forced to resign on 6 December 1990, and there were democratic elections in the next year.

Proponents of embankments saw floods as a problem and cited the Mississippi River in the US and dykes and polders in the Netherlands as models. Civil society protests against FAP pointed to importance of normal barsha floods as part of the agricultural system, to past failures of Western embankment attempts in Bangladesh such as Beel Dakatia and to breaches of embankments elsewhere such as the Mississippi in 1983 and the Netherlands in 1953.[23] One of the biggest problems with embankments was that the designers assumed that flooding was caused by the rivers, when in fact it is caused by heavy rainfall that cannot drain off because the rivers are in flood; this misunderstanding means that rainwater floods build up behind the embankments.[24]

Other than the Dhaka embankment, no other new embankment was ever built. FAP studies of previous flood protection projects confirmed grave weaknesses in design, management and maintenance.[25] In 1991 the World Bank's own Independent Evaluation Group (IEG) noted that it 'is now well known and accepted' that the Bank's 1978 Brahmaputra Right Embankment project 'does not give very good flood protection'.[26]

Faulty Justification

Even while the mountain of FAP studies were still being produced, the World Bank IEG pointed out that the underlying assumption of all of the embankment projects was faulty. The financial justification of 27 years of World Bank embankment projects had been entirely based on facilitating more intensive wet season agriculture and increased rice production, through surface water irrigation and protection against flood damage. Dykes were not designed to protect human lives or properties. Embankments, pumps and canals proved a very expensive and not very cost-effective way to increase food production. But IEG noted that the Bank had assumed 'that flood control would yield enough wet-season agricultural benefits to justify any engineering works that were technically feasible'.[27]

The IEG realized that a rapid and massive shift from monsoon rice to dry season irrigated *boro* rice, described in the next chapter, was already under way. The Bangladesh Rice Research Institute had produced high-yielding varieties of winter boro rice irrigated

23 Kelly Haggert et al. eds, *Rivers of Life* (Dhaka: Bangladesh Institute for Advanced Studies and London: Panos, 1994).
24 Hugh Brammer, 'After the Bangladesh Flood Action Plan: Looking to the Future', *Environmental Hazards* 9 (2010): 120.
25 Brammer, 'After the Bangladesh FAP', 118.
26 IEG, 'The World Bank Support for Flood Control'.
27 IEG, 'The World Bank Support for Flood Control' and Brammer, *Can Bangladesh be Protected?* 150–51.

from ground water with small pumps, which were being rapidly adopted by the farmers. Wet season rice was less important, and rice production was rising and prices were falling, which made a nonsense of the economic justifications of the embankments and polders. Instead of giant centrally managed irrigation systems, the IEG found that 'the fact that privately owned tube wells and low-lift pumps are now self-financing, self-operated and self-maintained indicated that farmers truly want minor irrigation equipment and that it is truly effective.'[28] World Bank support of engineering solutions and for FAP was over. The Bangladesh Water Development Board (BWDB) has always been the main promoter of engineering solutions; in the 1970s nearly 20 donors supported it but during the decade 2000 to 2010 it had only three donors supporting it.[29]

Flood Proofing

The 1995 Bangladesh Water and Flood Management Strategy called for selective structural interventions to protect strategic infrastructure and urban areas from river-bank erosion and floods, but not for rural areas, where flood proofing was advocated.[30] Flood proofing is similar to what is called 'adaptation' in climate change jargon. It involves risk reduction, early warning, and relief and reconstruction. The book *Climate Change Adaptation Actions in Bangladesh* by Rajib Shaw, Fuad Mallick and Aminul Islam provides more information on this subject.[31]

Bangladeshis have a set of risk reduction skills learned over generations, but honed by more recent experiences and academic research. For example, houses and villages are built on the highest available ground, often the narrow crests that mark the edges of old river channels; houses are often built on earth plinths or platforms. Climate change means higher floods, which means raising the level of houses, and of toilets and water pumps. The height of roads, railways and other key infrastructure must be raised above projected flood levels, and the number of bridges and culverts must be increased to ensure drainage. Estimate of the costs of these works before 2050 are more than $2 bn.[32]

Farmers have hugely complex management systems and show great flexibility, shifting their crops according to the start and end of rains and floods. Extra plants are grown for transplanted rice in case the first set is lost to drought or flood; short season crop seeds are kept in case a crop is lost to flood, so that some food can be grown in the moist ground. Even small farmers have more than one plot with different soil types and at different heights.[33]

One common practice is to plant seeds for aus and aman rice at the same time, with the aus harvested before the floodwaters are too deep and the aman harvested after the

28 IEG, 'The World Bank Support for Flood Control'.
29 Northwest Hydraulic, 'Bangladesh: Main River Flood', xvi
30 Brammer, *Can Bangladesh be Protected?* 172, 189, 210–11.
31 Rajib Shaw, Fuad Mallick and Aminul Islam, eds., *Climate Change Adaptation Actions in Bangladesh* (Tokyo: Springer, 2013).
32 Susmita Dasgupta et al., 'Climate Proofing Infrastructure in Bangladesh: The Incremental Cost of Limiting Future Flood Damage', *Journal of Environment & Development* 20 (2011), 183.
33 Brammer, *Can Bangladesh be Protected?* 137, 138, 141–47.

waters have receded, with an expectation that even in a bad year at least one of the two will be successful. Because the crops are planted from seed and not transplanted, the young plants must be thinned to obtain the correct spacing but ensure a mix of aus and aman. One of the authors remembers working in a field in Manikganj and being taught by his grandfather how to tell the difference between the small plants: there is a small circle just below the first leaf and it is purple on the aman but green on the aus. As we note in the next chapter, new rice varieties are being developed to support farmers in the flood proofing.

Several important aspects of FAP did go ahead. Flood forecasting and warning systems have been improved, and there are better relief and rehabilitation measures.[34] The BWDB has established the Flood Forecasting and Warning Center (FFWC) which operates in the flood season, from April to October. Its website (http://www.ffwc.gov.bd/) has rainfall and river levels and forecasts of river flooding five days into the future for 92 measuring stations. Use of radar altimetry satellite observations of river levels of the upper catchment water areas on the Ganges and Brahmaputra rivers has allowed flood forecast up to eight days ahead. Flood warning dissemination is also done through an interactive voice response (IVR) system using mobile phone number 10941. Flood information centres have been established at the division levels and they have links with local communities, NGOs and news media. NGOs disseminate flood warning messages generated by the FFWC at community level.

Modelling of river behaviour has also improved. For farmers, erosion is a much more serious problem than floods, and the government's Center for Environmental and Geographic Information Services (CEGIS) is now able to combine satellite images with river models to predict one year in advance areas of major riverbanks likely to be eroded.[35] But they are not always listened to, they say. In one example the World Bank was resettling people from a project along the Jamuna river, and CEGIS said the resettlement site was likely to be eroded; the Bank ignored the warning, but had to resettle the people a second time when erosion happened.

With generations of experience of flooding, Bangladeshis are immensely resilient and adaptable. But even they cannot cope with the destructive bonna floods, which destroy people's assets and livelihoods. When flood waters are deep there is no work for the urban rickshaw pullers and rural farm labourers, who often live from day to day and need each day's earnings to buy food. Amartya Sen, in his famous book *Poverty and Famines*, looked at the Bangladesh famine in which at least 26,000 people died after the huge floods in 1974, and concluded that the problem was not lack of food, but lack of 'entitlements' – in other words, poor people, especially landless labourers, lacked the money to buy food and there was insufficient food relief.[36] Thus the final part of flood proofing is the creation of relief

---

34 Brammer, 'After the Bangladesh FAP', 119.
35 Northwest Hydraulic, 'Bangladesh: Main River Flood', xii.
36 Amartya Sen, *Poverty and Famines: An Essay on Entitlement and Deprivation* (Oxford: Oxford University Press, 1982), Chapter 9, 'Famine in Bangladesh'. Sen (p. 136) does note that government did not have enough grain, in part because the United States withheld food aid because Bangladesh was exporting jute to Cuba.

and rehabilitation systems. First, it is necessary to provide adequate shelter, sanitation, water and food for those displaced. Second, financial assistance is necessary for people to rebuild their homes and to replace lost productive assets such as animals, tools and ploughs.

Learning and Innovation

Since independence, a growing group of universities and research institutes have been doing detailed studies of floods and responses, while local communities have become more active. The first response was for 'flood control' and imposing order on nature – an attempt to tame the rivers with dams, embankments and polders. By the mid-1990s there had been a paradigm shift, from flood control to a much broader sense of 'water management', which looked at floods, drainage, irrigation, navigation and environment. With respect to floods it takes a much more holistic view of floodplains and their role in storing water and replenishing groundwater.[37] This change has occurred with respect to both river floods and coastal flooding (Chapter 4).

Another paradigm shift has been from relief to risk reduction and comprehensive disaster management. Mahbuba Nasreen told us that after the 1987 and 1988 floods she realized that there was no academic research on disaster management – on how people actually cope with floods and what they know from practical experience. She has been part of a rapid shift wherein eight universities in Bangladesh now run courses on disaster management and Prof. Nasreen is now director of the Institute of Disaster Management at the University of Dhaka. She sees education as a way to train people to think about floods, and her institute is training secondary school teachers to train 10,000 primary teachers in disaster management. This links to the growing importance of flood proofing, where the risk to people and property is reduced, but where there is also a development of more effective and suitable immediate disaster protection, such as improved shelters, relief for those affected and the support to rebuild.

Women's education has been a priority for Bangladesh. Women's particular role in floods took longer to be recognized, but this is leading to another paradigm shift. Nasreen stresses that disasters 'affect both women and men but the burden of flood coping falls heavily on women. During floods men in rural areas lose their place of work while women shoulder the responsibilities to maintain households' sustenance'.[38] She told us that 'when the field goes under water, the men say they have no work to do.' It is women who do much of the flood proofing, including saving food stocks and fuel in dry places and lifting belongings to platforms. During the flood, women still have to provide clean water and food and treat illnesses such as diarrhoea, which are common during the

---

37 Rezaur Rahman and Mashfiqus Salehin, 'Flood Risks and Reduction Approaches in Bangladesh' in *Disaster Risk Reduction Approaches in Bangladesh*, ed. Rajib Shaw, Fuad Mallick and Aminul Islam (Tokyo: Springer, 2013), 80.

38 Mahbuba Nasreen, *Violence against Women During Flood and Post-Flood Situations in Bangladesh* (Dhaka: Action Aid, 2008), i.

high water.[39] Local communities are becoming more important in disaster management, but Nasreen argues that although women are on the committees, 'communities are not counting women's voices'.

The Bangladesh Bureau of Statistics (BBS) did a survey of violence against women in 2011, and found that 46 per cent of married women had experienced physical violence by the current husband in the past 12 months; 8 per cent of women had experienced violence from a non-partner during the previous year.[40] Prof. Nasreen researched on violence against women in floods in 2007 and found that violence was much worse during a flood. Women reported they had no protection against harassment and violence by neighbours and relatives during a flood.[41] She was shocked and told us, 'I had never even thought it was that bad.' But the fact that the BBS did a special survey of violence against women – the first of its kind in Bangladesh – is a small indication of change.

These paradigm shifts support the move to prepare for the impacts of climate change. They reflect an understanding that people cannot be completely protected from the depredations of climate change, but that study and understanding of previous disasters shows how to improve protection and response. More changes will be needed in coming years, but there is improvement and increased adaptation.

Conclusion: Adapting to Floods

Hugh Brammer's book title has the question *Can Bangladesh be Protected from Floods?* and his reply is 'the answer must be no,' at least not in the sense of the United States or Europe.[42] It would be much too expensive, and it is not something that the industrialized world has succeeded in doing, as illustrated by recent major floods of the Mississippi in 1993 and 2011 and in Europe of the Rhine in 1993 and 1995[43] and the Elbe and Danube in 2009 and 2013. In any case, the hydrology of Bangladesh is just too complex. This is a living delta with moving rivers and huge sediment flows, and people want normal barsha floods but protection from extreme bonna floods.

The outcome has different names – flood proofing, adaptation, living with floods, managing the floods – but they all mean a mix of individual, community and national actions. The Mughals of the seventeenth century improved existing channels to create a canal system, which managed and drained the monsoon floods. The British built railways and roads on embankments without enough bridges and culverts and the Pakistan and early post-independence period saw attempts at large-scale engineering solutions which proved to be correct to protect cities and to defend the coast against cyclone storm surges, but which caused huge problems for agriculture and the natural flooding systems.

39 Farhana Sultana, 'Living in Hazardous Waterscapes: Gendered Vulnerabilities and Experiences of Floods and Disasters', *Environmental Hazards*, 9 (2010): 43–53.

40 Bangladesh Bureau of Statistics (BBS), *Report on Violence against Women Survey 2011* (Dhaka: BBS, 2013), 30, 52.

41 Nasreen, *Violence against Women During Flood.*

42 Brammer, Can Bangladesh be Protected? 236.

43 Thomas Hofer and Bruno Messerli, *Floods in Bangladesh* (Tokyo: United Nations University Press, 2006), 431–32.

Protests stopped the FAP and communities developed Tidal River Management and are pushing for more openings in embankments and reversals of some of the failed engineering attempts.

Some engineering is required, including more and higher dykes to protect Dhaka and the coast, and climate change means they will need to be bigger and stronger. And there are questions. For example, should Chowhali upazila have been defended by stronger embankments, or is it not worth the cost?

But for most people, living with the more serious floods brought by climate change means more adaptation, and that means more learning, experimenting and improving flood proofing at all levels. Furthermore, as incomes and living standards rise, people will expect more protection. This, in turn, requires support from government and the international community to build stronger houses on higher plinths, to have a range of crops better adapted to increased rainfall and higher floods, and perhaps most importantly, to have relief and reconstruction systems in place to help people recover from the once-in-a decade destructive bonna floods.

Chapter Seven

AGRONOMISTS KEEPING AHEAD
OF CLIMATE CHANGE

Developments by national scientists have made Bangladesh self-sufficient in rice, and ongoing technical advances should keep ahead of climate change, at least until 2050. Two decades ago there were predictions that Bangladesh faced famine; instead there are rice exports and child nutrition has improved dramatically. This is the result of post-independence agricultural and social revolutions, which are also pointing the way to coping with climate change.

Rice is the staple grain and provides 75 per cent of the calories and 66 per cent of the protein of the average daily diet.[1] The huge variations in soil, climate and seasons in the country mean that an estimated 27,000 rice varieties have been developed by farmers over the centuries in Bangladesh. Some can grow 20 cm per day to stay above increasing water depths and can grow above 5 m of water;[2] others are adapted to using the pre-monsoon and monsoon rains. Farmers have seeds for several varieties for different seasons and rainfall patterns. There are three rice seasons (see Table 2.1) identified by the time of harvest – post-monsoon, pre-monsoon and during the monsoon. One has short and long growing varieties, leading to four common rice crops:

- *B Aman* is broadcast or directly seeded aman, planted in March/April under dry conditions but at the time of pre-monsoon showers in areas that normally have deep flooding. This is harvested in October/December.
- *T Aman* is transplanted aman, planted usually in July/August during the monsoon but where normal flood depth is less than 0.5 m. With a shorter growing time, it is also harvested in November/December.
- *Boro* is transplanted and grown under irrigation during the dry season. It is planted from December to March and harvested between April and June.
- *Aus* is sown during the pre-monsoon showers in March/April and is harvested in the monsoon season in June/July. Traditionally, it was broadcast sown (*B Aus*), but increasing amounts are now transplanted (*T Aus*) under irrigation on land that will not be

1   BRRI, 'BRRI at a Glance' (Gazipur: Bangladesh Rice Research Institute (BRRI), 2014), 2. Accessed 10 April 2016, http://brri.portal.gov.bd/sites/default/files/files/brri.portal. gov.bd/page/c8743e0a_87bf_46d6_9a16_bf6cd00db1c8/BRRI_at_a_Glance_2014.pdf. Another survey found 57 per cent of protein from rice, 18 per cent from fish and 5 per cent from lentils. Julia E. Heck et al., 'Protein and Amino Acid Intakes in a Rural Area of Bangladesh', *Food and Nutrition Bulletin* 31 (2010): 206–13.
2   Hugh Brammer, *Can Bangladesh Be Protected from Floods?* (Dhaka: University Press, 2004), 42.

deeply flooded in the monsoon season. *T Aus* is now considered as a late boro rice because it uses the same seeds as boro and is initially irrigated.

Bangladesh is hugely variable, but typically soils are moist enough to grow rice from April through November. Traditionally, most rice was aman, transplanted or seeded in the monsoon and grown in flooded fields. Next was aus, which used the rainfall of the pre-monsoon period to start growing. Much less important was boro dry season rice, which had to be irrigated, although traditional boro was also grown without irrigation in perennially wet depressions, especially in Sylhet. Many farmers try to grow two rice crops per year.

But all of the local varieties have relatively low yields, although they include some prized for their taste. The green revolution of the 1960s offered the opportunity for high-yielding varieties, which used fertilizer and pesticides to increase production. The International Rice Research Institute (IRRI) in the Philippines led the way with new high-yielding varieties based on rice from many countries, including East Pakistan. Two were introduced into India and Pakistan, IR-8 in 1967 and IR-20 in 1970.[3] Most local aman varieties were long-stemmed (greater than 1 m) to withstand floods, but nitrogen fertilizers increase the yield by increasing the size of the grain head, which meant they had the problem that they easily lodged (fell over). In response, IRRI varieties were short (known as dwarf varieties) with stiffer stems so as not to lodge, but could only be grown on non-flooded or very shallow flooded land in the aman season.

After independence national plant breeders at the Bangladesh Rice Research Institute (BRRI), working with IRRI, began to apply these modern breeding techniques in response to the needs of local farmers. Three main breeding techniques are used to produce new high-yielding varieties. Conventional or 'inbred' crop breeding involves combining different strains of a crop to gain desired traits and then continuing to grow them in fields until they are stable and each generation is the same. These crops are 'open pollinating' and farmers can keep the seeds and replant, as they traditionally do. Initial IRRI high-yielding rice was produced in this way. The second technique, developed in China and increasingly used by IRRI,[4] is crossbreeding to produce 'hybrid' rice in which the first generation is highly productive. But to gain this, new seeds have to be bought each year and farmers cannot use their rice grains as seeds. BRRI decided not to stress hybrid rice and mainly to produce inbred varieties, so farmers could keep their own seeds. However significant amounts of hybrid boro seed are now used. The third crop development technique is genetic modification (GM), which has so far not been successfully applied to rice.

---

3   One somewhat higher yielding variety was introduced pre-IRRI and gained some acceptance. It was Pajam (Pakistan-Japan-Malaysia) and was developed by Japanese researchers in Malaysia. It was eventually supplanted by higher yielding IRRI varieties. Yuko Tsujita, 'Introduction', in *Inclusive Growth and Development in India: Challenges for Underdeveloped Regions and the Underclass*, ed. Yuko Tsujita (Basingstoke, UK: Palgrave Macmillan, 2014), fn 8.

4   http://irri.org/our-work/research/better-rice-varieties/hybrid-rice, accessed 20 April 2016.

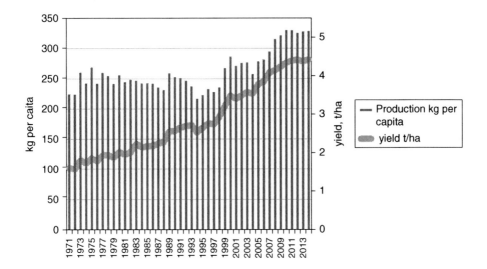

Figure 7.1 Per capita rice production and yield
Source: FAOstat, accessed 26 March 2016.

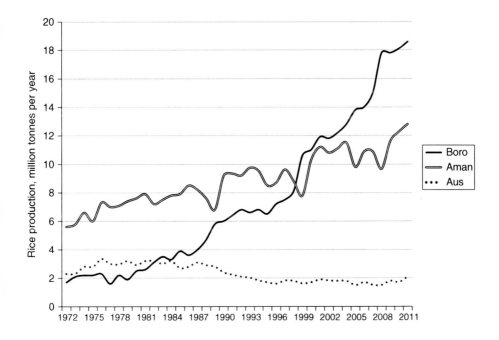

Figure 7.2 Rice production by season
Source: Akbar Hossain and Jaime A. Teixeira da Silva, 'Wheat and Rice, the Epicenter of Food Security in Bangladesh', *Songklanakarin Journal of Science and Technology* 35 (2013): 261–74.

After independence, BRRI began to release modern high-yielding rice varieties, and by 1980 these accounted for about a quarter of aus and aman rice production. The aman variety BR-11 was released in 1980 and remains popular; by 1992 more than half of aman rice was produced from modern varieties and this has continued to increase.[5] But the real change came in the 1980s with the introduction of modern varieties of irrigated winter boro rice, particularly BR-28 released in 1994.

Just producing new varieties is not enough. BR-28 was linked to improved agricultural extension and fertilizer subsidies, particularly for urea produced from Bangladeshi natural gas, plus a huge government irrigation programme and small scale mechanization. The Bangladesh Agricultural Development Corporation (owned by the Ministry of Agriculture) promoted the expansion of shallow tube well irrigation on relatively higher land, low-lift pumps on lower land near rivers and beels, and later deep tube wells. From the late 1980s the private sector played a major role. By 2010 there were 1.4 million shallow tube wells with pump sets, mainly with small Chinese diesel engines. There are also diesel low-lift pumps that take water from canals. The number of small two-wheel tractors (again, mainly Chinese) increased from 100,000 in 1996 to 400,000 in 2010 and to an estimated 600,000 more recently.[6]

In recent years new crops, irrigation and mechanization have been supported by cash subsidies to small farmers through an input distribution card that reduced the prices of fertilizer and for electricity and fuel for irrigation. For 2015, agriculture subsidies were $1.3 bn and food subsidies $230 mn.[7] In some areas, farmers purchase electronic cards with irrigation-hour credits and insert the cards into electronic terminals next to water pumps.[8] There have been important social and economic changes in rural areas. The rapid rise in rice production has increased the demand for labour, benefitting landless families, and rice farmers' income gains have been spent on non-farm goods and services, such as house improvements. Meanwhile, there is an increasing engagement of male labour in the non-farm sector, and in urban and foreign employment, which has resulted in more farms controlled by women. Female employment in agriculture jumped from 3.8 million in 2000 to 10.5 million in 2010.[9]

Since 2000 boro rice has accounted for more than half of all rice produced (see Figure 7.2). As Figure 7.1 shows, yields per hectare have risen steadily, but for the first

5   Akbar Hossain and Jaime A. Teixeira da Silva, 'Wheat and Rice, the Epicenter of Food Security in Bangladesh', *Songklanakarin Journal of Science and Technology* 35 (2013): 261–74.

6   Stephen Biggs and Scott Justice, *Rural and Agricultural Mechanization: A History of the Spread of Small Engines in Selected Asian Countries*, IFPRI Discussion Paper 01443 (Washington, DC: International Food Policy Research Institute, 2015), 5.

7   Rejaul Karim Byron, 'Subsidy to be Trimmed in Next Budget', *Daily Star*, 27 May 2015. Accessed 27 March 2016, http://www.thedailystar.net/business/subsidy-be-trimmed-next-budget-86887.

8   Stefanos Xenarios et al., 'Alleviating Climate Change Impacts in Rural Bangladesh through Efficient Agricultural Interventions', 2013. Accessed 10 September 2016, http://www.eldis.org/go/home&id=65099&type=Document#.VvkkqBIrLus.

9   M. A. Sattar Mandal, 'Agricultural Transformation in Bangladesh: Policies and Lessons Learned', presentation for panel discussion Agricultural Transformation in Asia, 23rd AERA annual conference, Mumbai, 2–4 December 2015.

two decades of independence this only kept pace with population growth. By the beginning of the twenty-first century, however, the falling birth rate (see Box 7.1) and the rapid adoption of boro rice meant that per capita rice production has been rising and by 2008 Bangladesh was self-sufficient in rice. Mechanization and irrigation has broader implications. Farmers already grow a wide range of crops, including potatoes, oilseeds, pulses, pumpkins, melons, tomatoes and other vegetables on tiny plots – the average farm size is less than 0.5 hectares. Irrigation has not just ensured higher, more reliable, rice production in the dry winter season, it has also raised production of some other irrigated winter crops which has made the country self-sufficient in potatoes, tomatoes and vegetables.[10]

Boro rice has a number of advantages. Most important, aus and aman rice are subject to the irregularities of rainfall and to flood damage, whereas water can be controlled for boro rice. Furthermore, most of Bangladesh has sufficient ground water for irrigation, and aquifers are recharged by the monsoon rains and floods. Aus rice has fallen out of fashion partly because it is hazard-prone and output is highly variable from year to year. Boro rice is more costly in terms of water, fertilizer and pesticides and needs closer management, noted M. Asaduzzaman, research director at the Bangladesh Institute of Development Studies.[11] Nevertheless, he argues, it is 'no wonder farmers wherever possible switch to irrigated boro to ensure food output'. Where two rice crops can be grown, they are usually boro and transplanted aman.

Since independence, BRRI has released 67 high-yielding varieties of rice, of which 63 are inbred and only 4 are hybrid. Seeds are sold by BRRI and its agents as well as by private seed companies. BRRI has also preserved germplasm from 8,000 rice varieties. It has also dealt with rice diseases and developed a range of new production technologies. By 2013, modern varieties accounted for 92 per cent of total rice production.[12] The stress on inbred varieties has proved wise – about half of the farmer's seed is saved from previous harvests, and half is purchased from shops. The larger farmers are more likely to use their own seed.[13]

After Boro

With boro rice well established and rice production rising, BRRI and other researchers have moved into a new set of overlapping areas. Stress tolerance, efficiency, and climate change have become priorities. BRRI is developing special rice varieties that

---

10 Mohammed Mainuddin et al., 'Spatial and Temporal Variations of, and the Impact of Climate Change on, the Dry Season Crop Irrigation Requirements in Bangladesh,' *Irrigation Science*, 33, (2015): 116.

11 M. Asaduzzaman, 'Technology Transfers and Diffusion: Simple to Talk About, Not So Easy to Implement,' talk given at WIPO Conference on Innovation and Climate Change, Geneva, 11–12 July 2011.

12 BRRI, 'BRRI at a Glance'.

13 Akhter U. Ahmed et al., *The Status of Food Security in the Feed the Future Zone and Other Regions of Bangladesh: Results from the 2011–2012 Bangladesh Integrated Household Survey* (Dhaka: International Food Policy Research Institute, report for USAID, 2013), 140, table 4.14.

better resist various stresses. For submergence-prone areas there are varieties that are flash-flood resistant. Some are cold or salt tolerant. Between 1960 and 1991 there were 19 droughts somewhere in Bangladesh, and of those seven were severe and affected significant parts of the country. For drought areas BRRI has rice varieties that can survive longer without water and others that mature more quickly and can be harvested a month earlier than other rice varieties and avoid drought stress. Increased irrigation plays a role during drought.

More attention is being paid to the special problems of farmers in specific areas. Soil and ground water salinity is increasing in some coastal areas and a recent study showed an increase in soil salinity between 2001 and 2009 in all 41 measuring stations.[14] 'Promoting adaptation to coastal crop agriculture to combat increased salinity' was presented as the number one adaptation intervention in the 2005 National Adaptation Programme of Action.[15] A salt-tolerant boro rice variety BR 47 has been introduced in coastal areas, permitting double cropping in those zones.[16] The World Bank points to the need to increase surface flows of water from upstream; this partly involved tidal river management (as noted in Chapters 5 and 6) and better operation of sluices and regulators.[17]

In flooded or waterlogged areas, particularly in the south, a traditional practice of floating gardens is being expanded. Farmers create 2 m × 10 m platforms from local materials – water hyacinth, rice stubble, coconut husk and other plant waste held together by bamboo. The water hyacinth rots over a month to create organic manure and the floating platform becomes a hydroponic garden for vegetables and beans.[18]

Boro production has increased because irrigation allows a more regular water supply, compared to the irregular monsoon. Thus aus production has been falling and aman production rising only slowly, but BRRI is working on high-yielding modern varieties that can survive more variable rain and flood. BRRI is also looking at increased mechanization for those crops.

The next move will be towards more efficient production systems. One BRRI success is the urea deep-placement technology for increasing nitrogen-use efficiency, and which involves the placement of urea supergranules or briquettes at 7–10 cm soil depth a week after transplanting. This has significantly decreased fertilizer consumption while

14 Susmita Dasgupta et al., 'Climate Change and Soil Salinity: The Case of Coastal Bangladesh', *Ambio* 44 (2015): 815–26.

15 Ministry of Environment and Forest, *National Adaptation Programme of Action* (2005), 22. Accessed 28 March 2016, http://unfccc.int/resource/docs/napa/ban01.pdf.

16 Mesbahul Alam et al. 'Coastal Livelihood Adaptation in Changing Climate: Bangladesh Experience of NAPA Priority Project Implementation', in *Climate Change Adaptation Actions in Bangladesh*, ed. Rajib Shaw, Fuad Mallick and Aminul Islam (Tokyo: Springer, 2013), 263.

17 World Bank, *Bangladesh Climate Change and Sustainable Development*, report 21104-BD (Washington: World Bank, 2000), xvi.

18 Md. Anwarul Abedin and Rajib Shaw, 'Agriculture Adaptation in Coastal Zone of Bangladesh', and Mesbahul Alam et al., 'Coastal Livelihood Adaptation in Changing Climate: Bangladesh Experience of NAPA Priority Project Implementation', in *Climate Change Adaptation Actions in Bangladesh*, ed. Rajib Shaw, Fuad Mallick and Aminul Islam (Tokyo: Springer, 2013), 221, 267.

increasing yield. Another is a simple leaf colour chart which is a sheet of plastic with four shades of green, from yellowish to dark green; comparing the chart to the rice leaves at regular intervals tells the farmer if urea has been under- or over-applied. Very simple but very effective.

Using Less Water

There has been a huge increase in irrigation, from just 18 per cent of cultivable area in 1982 to 63 per cent in 2010. Now, changing irrigation systems to use less water has become an important issue. There are four reasons. Two are straightforward. First, using less water means less pumping, which means using less fuel for pumps, saving money and greenhouse gases. Second, in some areas, such as the Barind Tract in the northwest bordering India, and Khulna in the south, there are indications of shortages of groundwater, which limits the expansion of boro rice with current irrigation methods.

The other two reasons are arsenic and methane, each of which has a longer and more complicated story. Arsenic in water was an unexpected result of trying to provide safe drinking water, but also has implications for irrigation. Many children suffered diarrhoea and disease from contaminated water sources, and thus the widespread introduction of 'safe' water from tube wells in the 1970s and 1980s in a UNICEF programme was considered a major public health success. But in the 1990s it was discovered that the water from many of the wells was contaminated with naturally occurring inorganic arsenic. Arsenic has long-term effects including skin lesions, cancers and neurological damage, and in 2000 arsenic in Bangladesh water was called 'the largest mass poisoning of a population in history'.[19] In 2010 it was estimated that 15 per cent of the population were drinking water with arsenic above the Bangladesh standard (50 µg/l) and 24 per cent of people were above the WHO guideline (10 µg/l).[20] Arsenic occurs naturally and was deposited ten to twelve thousand years ago after the last ice age when melting glaciers carried huge amounts of sediment from the Himalayas, which were later buried by new deposits.[21] Arsenic occurs in water in many countries, including Argentina, Spain and the United States; in southwestern England some private water supplies are over the WHO guideline. In Bangladesh the distribution of arsenic is quite variable, sometimes being different from one side of a village to the other. Over the past decade there has been a major programme to close heavily contaminated drinking water wells and find alternative water sources, including deeper wells, filter systems and rain water collection. But irrigation water remains a problem and irrigated rice accounts for more than half of the

19   Allan H. Smith, Elena O. Lingas and Mahfuzar Rahman, 'Contamination of Drinking-Water by Arsenic in Bangladesh: A Public Health Emergency', *Bulletin of the World Health Organization*, 78 (2000): 1093.

20   Dipankar Chakraborti et al., 'Status of Groundwater Arsenic Contamination in Bangladesh: A 14-Year Study Report' *Water Research* 44 (2010): 5789 and Dipankar Chakraborti et al., 'Groundwater Arsenic Contamination in Bangladesh – 21 Years of Research', *Journal of Trace Elements in Medicine and Biology* 31 (2015): 237–48.

21   Hugh Brammer, *The Physical Geography of Bangladesh* (Dhaka: University Press, 2012), 87.

arsenic consumed by people in food.[22] It is not practical to filter irrigation water, although alternative wells are sometimes possible.

Both arsenic and shortages of ground water occur only in some parts of Bangladesh. But in both cases there is a need to reduce the amount of irrigation water used and to try and use more rain and river water.

The fourth issues is methane (CH_4), which is produced by anaerobic[23] decomposition of organic material in flooded rice fields. The IPCC 2007 *Fourth Assessment* estimated that rice accounts for 17 per cent of global anthropogenic (human) methane emissions, compared to 26 per cent from cattle and other ruminants.[24] Farmers normally aim to keep fields flooded or wet up to the crop grain-filling stage, about 4–6 weeks before harvesting. Since, for boro, this is in the pre-monsoon period of heavy showers, irrigation may only be necessary for this stage in some years. But if irrigation is necessary, farmers usually use 'flood irrigation', trying to maintain water at a depth of 5–10 cm. Flooding is used both for weed control, and because rice was grown in naturally flooded monsoon conditions and it can tolerate wet soil conditions, which most other crops cannot.

A view that is increasingly gaining acceptance is that reducing the anaerobic conditions that produce methane also improves root growth and supports the growth of useful aerobic soil organisms, and thus increases rice productivity. This leads to the idea that rice field soils should be kept moist rather than continuously saturated by flood irrigation, which is often linked to various conservation agriculture techniques now being developed and which usually also involves reduced tillage or ploughing and reducing weed growth through mulch or ground cover crops. Using less flood irrigation is behind both Bangladesh's own alternate wetting and drying (AWD) system, in which the surface soil is allowed to dry out before the field is irrigated again, and the System of Rice Intensification (SRI), which comes originally from Madagascar. With AWD fields are alternately flooded to a depth of 5cm and then allowed to dry out until the water level in the soil (as measured in a cylinder placed in the soil) falls to 15cm below the surface. A BRRI study of AWD showed that putting on water three days after the disappearance of the previous irrigation water used 22 per cent less water but produced 5 per cent more rice.[25] AWD techniques can reduce methane emissions by 60 per cent.[26] Under SRI,

22 Md. Kawser Ahmed et al., 'A Comprehensive Assessment of Arsenic in Commonly Consumed Foodstuffs to Evaluate the Potential Health Risk in Bangladesh', *Science of the Total Environment* 544 (2016): 125.

23 Anaerobic is without air or without oxygen. Certain chemical processes require oxygen, but others can only take place when oxygen is not present. The water in flooded rice fields fills the spaces in the soil, so there is no air, and anaerobic processes can take place.

24 S. Solomon et al., eds, *Contribution of Working Group I to the Fourth Assessment Report of the Intergovernmental Panel on Climate Change* (Cambridge, UK and New York: Cambridge University Press, 2007) ch. 7, '7.4.1 Methane', average of values in table 7.6. Accessed 10 April 2016, https://www.ipcc.ch/publications_and_data/ar4/wg1/en/ch7s7-4-1.html#table-7-6.

25 M. Maniruzzaman et al., 'Validation of the AquaCrop Model for Irrigated Rice Production under Varied Water Regimes in Bangladesh', *Agricultural Water Management* 159 (2015): 331–40.

26 Xiaoyuan Yan et al., 'Statistical Analysis of the Major Variables Controlling Methane Emission from Rice Fields', *Global Change Biology* 11 (2005): 1131–41.

rice is grown and irrigated as a dry-land crop, on raised beds for drainage if necessary. SRI reduces water consumption by one-third. In its 'Intended Nationally Determined Contributions' submitted before the Paris COP in 2015, Bangladesh proposed to 'scale up rice cultivation using alternate wetting and drying irrigation' as one way to reduce greenhouse gas emissions. By 2030 it wants 20 per cent of all rice cultivation to use alternate wetting and drying.[27]

Perhaps the main argument against these techniques is that flooding is important for weed control and SRI, and conservation agriculture in general, usually requires more weeding and thus more labour, at least initially. These techniques are being studied and introduced, but are not yet widely adopted in Bangladesh. SRI remains controversial and led to an angry exchange in *The Geographical Journal*, in which the advocates of SRI were called 'highly misleading'.[28] Whatever the outcome of the specific SRI debate, Bangladesh will need to move towards using less water for rice.

Conclusion: Rice OK until 2050, at Least

'Bangladesh has recently achieved self-sufficiency in rice, due mainly to increased yields and the greatly increased area of groundwater irrigated dry season rice over the last several decades,' note Mohammed Mainuddin and Mac Kirby of The Commonwealth Scientific and Industrial Research Organisation (CSIRO).[29] They conclude that 'rice yields are well below potential and the current trends in yield, combined with the continued development of higher yielding varieties and more productive management practices, should enable Bangladesh to remain self sufficient in rice at least to 2050.' This is due to decades of work by Bangladeshi agronomists, producing new varieties of rice and improved farming methods. A country once dependent on food aid is now exporting rice. Research is already underway to raise productivity and to use less irrigation water. As Khorshed Alam in an article in the journal *Agricultural Water Management* concluded, Bangladeshi 'farmers are inherently resilient to a changing climate'.[30] But climate change will bring worse floods and more variable rainfall, which requires a redoubling of efforts by the agronomists.

---

27 Ministry of Environment and Forests (MOEF), 'Intended Nationally Determined Contributions', 2015, 7. Accessed 14 April 2016, http://www4.unfccc.int/submissions/INDC/Published%20Documents/Bangladesh/1/INDC_2015_of_Bangladesh.pdf.

28 Amir Kassam and Hugh Brammer, 'Combining Sustainable Agricultural Production with Economic and Environmental Benefits', *The Geographical Journal* 179 (2013): 11–18; James Sumberg et al., 'Response to "Combining Sustainable Agricultural Production with Economic and Environmental Benefits."' *The Geographical Journal* 179 (2013): 183–85; and Amir Kassam and Hugh Brammer, 'Reply to Sumberg et al.', *The Geographical Journal* 179 (2013): 186–87.

29 Mohammed Mainuddin and Mac Kirby, 'National Food Security in Bangladesh to 2050', *Food Security* 7 (2015): 633–46.

30 Khorshed Alam, 'Farmers' Adaptation to Water Scarcity in Drought-Prone Environments: A Case Study of Rajshahi District, Bangladesh', *Agricultural Water Management* 148 (2015): 205.

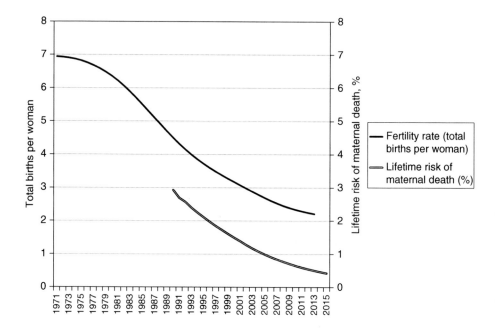

Figure 7.3 Birth rate and maternal mortality
Source: World Bank: World Development Indicators, accessed 28 March 2016.

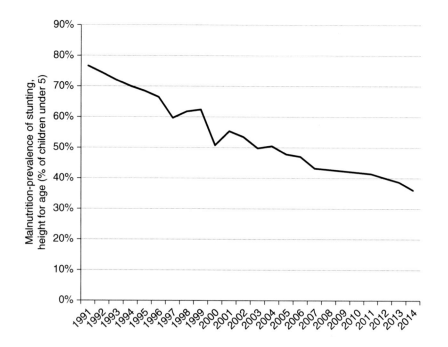

Figure 7.4 Child malnutrition
Source: World Bank: World Development Indicators, accessed 26 March 2016.

Box 7.1 Fewer but healthier children change the food balance

Older predictions of desperate food shortages in Bangladesh to be made worse by climate change were based on assumptions of a population explosion, which has not happened. Instead the birth rate has fallen quickly and dramatically. At independence, the average woman had 7 children; now it is down to 2.2. This is a result of rapid expansions of women's education and employment, sanitation, and health and family planning services. Together with increased food production, this has an impact on child nutrition and survival. Infant mortality has dropped from 15 per cent of children dying before they were one year old at independence in 1971 to only 3 per cent in 2015.[31] The International Food Policy Research Institute did a statistical analysis to find the causes of the rapid drop in child malnutrition and found three key factors – increased rice production was most important, but the two other important factors were increased education of mothers and improved household sanitation (notably toilets).[32] The end to rapid population growth changes the balance and makes Bangladesh more able to confront climate change.

---

31 World Bank, World Development Indicators, accessed 25 March 2016. http://data.world-bank.org/.

32 Derek D. Headey and John Hoddinott, *Agriculture, Nutrition, and the Green Revolution in Bangladesh*, IFPRI Discussion Paper 01423 (Washington, DC: International Food Policy Research Institute, 2015).

Chapter Eight

NO CLIMATE CHANGE MIGRANTS – YET

'Climate Change Makes Refugees in Bangladesh – Bangladesh and Countries like It Are on the Frontline of Mass Migrations as a Result of Global Warming' headlined *Scientific American* in 2009.[1] However most of the people interviewed by author Lisa Friedman came from the southwest and had been forced to migrate due to waterlogging caused by the construction of dykes and polders (see Chapter 4) or due to cyclone Sidr.

There are no 'climate change refugees' in Bangladesh, because climate change has not yet had a noticeable impact and because the Bangladesh climate is so variable. There is a whole book reporting on temperature and rainfall change which finds that over 50 years there had been a slight decrease in winter temperatures in the northwest, and elsewhere there had been a slight increase. These changes are real but not yet noticeable. For example, in the northwest the year-to-year variation in the maximum winter temperature can be 2°C or more; long-term data shows there is a decline in the maximum winter temperature in the northwest of 0.01–0.03°C per year, which is too small to be noticed.[2]

Scientists and statisticians can already see the effects of climate change, but ordinary people do not, and it does not have any effect on a decision to migrate. Yet. But this country's hugely variable climate means the norm is cyclones, floods, heavy rain and droughts. As discussed in Chapter 2, climate change will have major impacts. The national census shows that 80 per cent of those who migrate to Dhaka do so to look for work or because of poverty; only 8 per cent move directly for environmental reasons (see Box 8.1). However, poverty and unemployment is often caused indirectly by environmental factors – floods and cyclones destroy houses, productive assets and jobs; when these cannot be replaced, people migrate. We use the term 'environmental migrants' for people who move, directly or indirectly, due to cyclone, flood, erosion and waterlogging. Bangladesh has had environmental migrants for centuries, although the number is probably rising as population growth pushes people to live in more marginal areas.

---

1   Lisa Friedman, 'Climate Change Makes Refugees in Bangladesh', *Scientific American*, 3 March 2009 and further articles 16, 23 and 31 March 2009. Accessed 17 March 2016, http://www.scientificamerican.com/article/climate-change-refugees-bangladesh.

2   Tawhidul Islam and Ananta Neelim, *Climate Change in Bangladesh* (Dhaka: University Press, 2010), 34, 66.

Box 8.1 Dhaka slum arrivals

Reasons for coming to slums[3]

Seeking job	51%
Poverty	29%
River erosion and natural calamities	8%
Other	12%

When you last moved, did anyone help you to find a place?[4]

No one	33%
Family or relative	50%
Friend or contact	17%

Climate change is expected to increase the number of environmental migrants, but it has not done so yet, which creates a problem and a paradox for Bangladesh. Donor agencies want to help climate change migrants now, to show that aid is already helping. Journalists want climate change migrants because it makes their articles look better. Bangladesh needs the publicity and wants the money, so environmental migrants are re-labelled climate change migrants. In recognizing just how serious climate change will be later in this century, there is a tacit agreement to exaggerate what has happened so far. A similar problem occurs with researchers. One study set out to look for environmental migrants and asked 'what were the dominant hazards that compelled you to migrate?' With 'hazard' in the question, it is not surprising they then concluded that 'an overwhelming majority' were environmental migrants.[5]

In contrast, the Bangladesh Bureau of Statistics *Census of Slum Areas* asked the more open question of 'reasons of coming to slum areas'. The results in Box 8.1 are primarily economic and only 8 per cent environmental. An international group of researchers, half of it from the BRAC University Centre for Global Change in Dhaka, say, 'We conclude that, instead of climate change, social inequality and food insecurity as well as structural

---

3 Bangladesh Bureau of Statistics (BBS), *Preliminary Report on Census of Slum Areas and Floating Population 2014* (Dhaka: BBS, 2015). Accessed 24 April 2016, http://www.bbs.gov.bd/WebTestApplication/userfiles/Image/Slum/Preli_Slum_Census.pdf.

4 David Hulme, Manoj Roy, Ferdous Jahan and Simon Guy, 'Community and Institutional Responses to the Challenges Facing Poor Urban People in an era of Global Warming in Bangladesh', University of Manchester in collaboration with BRAC University, household survey, 2013, unpublished data.

5 Mohammad Abdul Munim Joarder and Paul W. Miller, 'Factors Affecting Whether Environmental Migration is Temporary or Permanent: Evidence from Bangladesh', *Global Environmental Change* 23 (2013): 1511–24.

economic differences are the strongest drivers of migration inside Bangladesh.'[6] Most researchers would agree with that.

It is useful to distinguish three groups of people who move:

- displaced people forced to flee a natural event such as flood;
- temporary migrants who travel to find work, such as farm labouring, for days or a few weeks; and
- permanent migrants who move permanently to a new location because of lack of work or those looking for a better job.

Many rural families, especially the landless and those with very small pieces of land, use temporary migration and seasonal labour as a livelihood strategy and to increase family income. Many seasonal migrants have established links with labour brokers and move when there is a temporary high demand for labour.[7] Other seasonal migrants have customary relationships, going to the same family every year, particularly for planting and harvesting crops in areas with big land holdings. People displaced by flood, cyclone or erosion normally move a short distance, usually to the nearest safe village, and hope to return home and rebuild. They are dependent on assistance from government or NGOs, or from better-off families who still see it as their social responsibility.

Migration is often reluctant – people try to stay as close to home as possible, although mobile telephones and new technology make it easier to keep in contact with families, and this does seem to facilitate migration. Men sometimes migrate in order to earn more money, leaving families behind. A study in Kurigram in northern Bangladesh showed that 43 per cent of households had members not present due to temporary or permanent migration, and that one-fifth of households rely on remittances from migrant family members in order to buy sufficient food.[8] It is also noted that even where migrants do not remit money, remaining family members gain because there are fewer mouths to feed.[9] Women who are unmarried, widowed or divorced sometimes migrate to the city because there are jobs in the garment factories.[10] This is a major social change as the rapid growth of garment factories has created large numbers of jobs for women.

Kurigram is a drought-prone district, and interviews in Khanpara village found important social differentiation in migration and identified four social groups.[11] A comparatively

6 Benjamin Etzold et al., 'Clouds Gather in the Sky, but no Rain Falls. Vulnerability to Rainfall Variability and Food Insecurity in Northern Bangladesh and its Effects on Migration', *Climate and Development*, 6 (2014): 18. doi: 10.1080/17565529.2013.833078.

7 Ibid., 23.

8 Ibid., 22.

9 Katha Kartiki, 'Climate Change and Migration: A Case Study from Rural Bangladesh', *Gender & Development* 19 (2011): 23–38. Accessed 10 September 2016, http://dx.doi.org/10.1080/13552074.2011.554017.

10 Matthew Walsham, *Assessing the Evidence: Environment, Climate Change and Migration in Bangladesh* (Dhaka: International Organization for Migration, 2010), 14.

11 Etzold, 'Clouds Gather', 23–25.

rich group with larger farms and non-farm income thus does not need to migrate to ensure income. Nonetheless, 31 per cent of those families included migrants, mainly men who moved for education or formal employment. Of middle families dependent on rainfed agriculture, 41 per cent had migrants, with a mix of permanent urban migrants and temporary rural ones. In the poor group, 49 per cent of families had migrants, most as temporary agricultural labourers or working in urban areas as rickshaw pullers or garment labourers. This third group is poor families who do not grow enough to meet their own food requirements and depend on day labour to earn money. Because there is little work during the 'monga' or hungry season, mid-September to mid-November just before harvest of aman rice in that area, many become temporary migrants during or shortly after the monga season. Finally, the fourth group is extremely poor landless households, who have few migrants because, as the study notes, 'they neither have adult male family members who could work as labour migrants, nor the required resources to facilitate migration, nor access to the necessary migration networks.'

This points to two important issues relating to response to disaster. First, farmers lose production and sometimes assets and migrate to replace that, and thus migration is temporary, while landless labourers lose their work and migrate to find other work and their migration may become permanent. Second, even temporary migration requires organization and money. Many temporary migrants gain work through labour brokers and often develop longer-term relations with those brokers who provide work in times of greatest labour demand. All studies show that most urban migrants already have family or other contacts that facilitate their arrival in the city, or have contacts through previous temporary migration. Katha Kartiki notes that 'households that lack financial and social capital are unlikely to undertake long-distance migration because of the costs involved'.[12]

Most studies show no clear link between disaster displacement and permanent migration – people initially hope to return home and rebuild their houses and productive assets, with the assistance of NGOs, government and temporary migration. One study notes that 'flooding undoubtedly causes substantial short-term population displacement, but it appears that this translates into few long-term moves'.[13] Even erosion migrants who have lost their land hope to resettle nearby. It is poverty, food insecurity and perceived risk that lead to a later decision to migrate permanently. That can be exacerbated by the environmental disaster, but the link is not direct or immediate. Indeed, natural disasters can potentially reduce migration by removing access to the resources needed to migrate – and also because the reconstruction can create local jobs and post-disaster relief can encourage people to stay.[14]

Personal contacts are key to migration. The first person in a village to migrate is brave and on their own, but in this era of mobile telephones the next migrants can make contact in the city, have a place to stay, have help to find work and become integrated. This is important in becoming established in semi-formal work, such as rickshaw pulling, which

12 Kartiki, 'Climate Change and Migration', 31.
13 Clark L. Gray and Valerie Mueller, 'Natural Disasters and Population Mobility in Bangladesh', *Proceedings of the National Academy of Sciences* 109 (2012): 6004.
14 Gray and Mueller, 'Natural Disasters', 6000, 6003.

is the main first employment for migrants. Family and village networks are important and two-way, because urban migrants often send money back to their village, helping their community to survive. Thus migration is not usually done in panic, but is planned and organized. The process is also complex with different groups acting differently. Some families in coastal areas send their children to Khulna for secondary education, which is a valuable investment in any case and also provides a base in the city if the need to migrate arises, even temporarily.[15]

This, in turn, points to the complexities of migration. Using their contacts, some people may try migrating to several different places before they settle. At the other end of the spectrum, especially in Khulna, which is only two hours by public transport from some coastal areas, people fled to Khulna during storms with the expectation of at least part of the family returning to their farms. Some divided families can maintain a long-term relationship between the Khulna part and the rural part of the extended family. But waterlogging, unrepaired broken dykes, eviction by thugs and other causes sometimes make it difficult to return, and households slowly become permanent residents of Khulna slums.[16]

Environmental Migrants

Environmental factors are not the immediate cause of people permanently migrating to cities, and climate change has, so far, had no impact. But the impact of environmental events, from cyclones to floods to waterlogging, has a direct effect on people's ability to continue living in rural areas. Thus even if the main reason to migrate is to obtain a better job or escape poverty, many of these people have been pushed by environmental problems.

Erosion can be quite dramatic, eating away large sections of river bank in a single monsoon season. For example, the 220 km long Jamuna River (the downstream part of the Brahmaputra) in the period 1984–92 widened by 184 m per year, eating 100 m annually from the left bank and 84 m along the right bank. This meant 5,000 ha/year was lost to erosion – although some of this soil was deposited in the river as islands, know as chars, at a rate of 900 ha/year.[17] Families in the area treat erosion as normal and move quite short distances, on average only 2.2 km, but move several times in their lifetime.[18]

Similarly, victims of floods, cyclones and storm surges try to return to their homes and rebuild. In waterlogged areas of the southwest, the first step is to try to find other local income sources.

---

15 Manoj Roy, Ferdous Jahan and David Hulme, 'Community and Institutional Responses to the Challenges Facing Poor Urban People in Khulna, Bangladesh in an Era of Climate Change', BWPI Working Paper 163 (Manchester, UK: Brooks World Poverty Institute, University of Manchester, 2012), 34.

16 Ibid., 46.

17 M. R. Rahman, 'Impact of Riverbank Erosion Hazard in the Jamuna Floodplain Areas in Bangladesh', *Journal of Science Foundation* 8 (2010): 55–65. (The journal of the Bangladesh Science Foundation.)

18 C. E. Haque and M. Q. Zaman, 'Coping with Riverbank Erosion Hazard and Displacement in Bangladesh: Survival Strategies and Adjustments', *Disasters* 13 (1989): 300–14.

Environmental stresses also change the local economic balance. For example, salinization in the southwest encourages the move from rice to shrimp farming. Shrimp have become the fourth most important source of foreign exchange – $440 mn in 2014–15. It has grown rapidly since the 1970s, partly promoted by the World Bank. Shrimp culture is mainly done in a traditional way and annual yield is just 250–300 kg/ha. The total shrimp farming area is 276,000 ha, mainly in the southwest coastal zone.[19] Shrimp and fish have traditionally been cultivated in canals and flooded fields, but large-scale commercial cultivation has been controversial because it reduces the number of people employed on the land. Processing the shrimp creates jobs in Khulna and elsewhere, which encourages people to migrate, although the shrimp processing industry operates with tight syndicates and shrimp cutters and cleaners work in closed groups – so family links or some other entry is essential. But the extra processing jobs are fewer than the farm jobs lost, so the conversion of rice fields to shrimp ponds is often opposed. Shrimp live in brackish water, so shrimp farmers sometimes cut the dykes to let in salt water, which also damages neighbouring rice fields. Shrimp farmers tend to be better off and more influential, and sometimes use musclemen to intimidate farmers to give up their land for shrimp.[20] Some small farmers have used microcredit to start their own shrimp farms, but the costs and risks are high. The issue remains extremely controversial. The World Bank, which initially promoted shrimp culture, also warned about 'its adverse environmental and ecological effects and serous social problems'.[21] The government's 2013 *Master Plan for Agricultural Development in the Southern Region of Bangladesh* says shrimp farming should be promoted along with attempts to increase yield, which is still low.[22] So shrimp farming continues to expand and forces landless people and those with small farms to migrate to find work. These people could be seen as environmental migrants, but in practice they are moving because of economic change, not climate change.

A study in Satkhira in the coastal southwest of Bangladesh showed significant migration, particularly to neighbouring India. In the two unions studied, Gabura and Munshiganj, after cyclone Aila in 2009, 10 per cent of the population migrated temporarily or permanently.[23] The main reason for migration was 'livelihood insecurity'

---

19 S. Humayun Kabir, 'Sea Food Export from Bangladesh and Current Status of Traceability'. Accessed 24 March 2016, http://www.unescap.org/sites/default/files/6-%20%20Sea%20 Food%20Export%20from%20Bangladesh-Kabir.pdf.Note that the words shrimp and prawns are used confusingly and interchangeably. In Britain and the Commonwealth, prawns are larger than shrimp, while in the United States shrimp is a more general term. Some Bangladesh Fisheries Ministry reports simply use shrimp, while others distinguish between 'bagda', which are identified as 'black tiger shrimp' and 'golda', which are 'giant freshwater prawns'.

20 Roy, Jahan and Hulme, 'Community', 20.

21 World Bank, *Bangladesh Climate Change and Sustainable Development*, report 21104-BD (Washington, DC: World Bank, 2000), 19.

22 Ministry of Agriculture, *Master Plan for Agricultural Development in the Southern Region of Bangladesh* (Dhaka: UN Food and Agriculture Organization, 2013), 59.

23 Ainun Nishat et al., 'Is Livelihood Insecurity Resulting from Impacts of Climate Change Leading to Enhancement in Migration to Urban Areas in the Coastal Region of Bangladesh?' paper presented by Nandan Mukherjee (C3ER, BRAC University) at the "ClimbUrb International Workshop", University of Manchester, UK, 9–10 September 2013.

caused by flooding and damage due to Aila; waterlogging and a transformation of agricultural land to shrimp farming were also important factors. We can see these people from Satkhira as environmental migrants, driven indirectly by cyclone Aila and shrimp farming, but not yet as climate change migrants. Analysis of Satkhira weather records show what is predicted for initial phases of climate change, with little overall change but more extreme events. However, so far the changes are almost unnoticeably small. Over 70 years, rainfall has not increased and temperature is rising by only 0.014°C per year. But there is a statistically significant increase of rain on days with heavy rainfall, and a significant increase of days with high temperatures.

More local intervention to support environment-affected people could cut migration. One study showed that of environmentally displaced people in Khulna, none had received government or NGO support before they left their village.[24] Delayed dyke repairs after cyclone Aila meant many people never returned. Intervention to help people to stay in rural areas and not migrate would require three phases: 1) after a natural disaster, temporary shelter, food, water and toilets; 2) support to rebuild both homes and livelihoods, including replacing lost animals and equipment; and 3) assistance to bridge income gaps, through work – perhaps as part of any repair and rehabilitation projects – or cash transfers. Climate change will increase damage and disruption, and thus make intervention more important if migration is to be avoided.

International Migration

Ten million Bangladeshis work aboard and send more than $15 bn back to Bangladesh each year.[25] As the government admits, 'Almost two million additional young people are added to the labour force every year, and the country lacks the ability to create jobs to accommodate all of them. As a result, the outflow of Bangladeshi workers will continue in the foreseeable future.'[26] In contrast, Bangladeshi garment exports were $25 bn in 2014/15, so exporting people is almost as important as exporting clothing.

International migration is a long-term affair. The Bangladesh Bureau of Statistics (BBS)[27] reports that 35 per cent of migrants have been away for 5–9 years and 23 per cent for 10 years or more. Half of remittances come from the Gulf states. The other two important countries are the United States and Malaysia.[28] One important group is educated people who have studied abroad and who remain in Europe or North America, but continue to send remittances to their families.

24 Roy, Jahan and Hulme, 'Community', 55.
25 'Bangladesh Remittances'. Accessed 18 March 2016, http://www.tradingeconomics.com/bangladesh/remittances.
26 Bangladesh Bureau of Statistics (BBS), *Survey on the Use of Remittance 2013* (Dhaka: Bangladesh Bureau of Statistics, Ministry of Planning, 2014), xix.
27 BBS, *Survey on the Use of Remittance*, 25
28 Bangladesh Bank (Central bank), 'Wage Earners Remittance inflows: Country wise'. Accessed 18 March 2016, https://www.bb.org.bd/econdata/wagermidtl.php.

BBS[29] finds that remittances are 78 per cent of the total earnings of remittance-receiving households.

Despite the fact that agriculture is the largest source of non-remittance income for these households, more than half of them have less than 0.2 hectares of land. 'This is a pointer to the overall economic hardship' of these households, BBS notes. 'The above findings imply that migration as strategy to earn remittance is still adopted pre-dominantly by low income group.' One-third of remittance-receiving households are in Dhaka, one-third in Chittagong, and one-third elsewhere.

Housing investments account for nearly half of remittances: purchase of land, 15 per cent; purchase of flats, 5 per cent; and construction of houses, 24 per cent. There is little productive investment, with fewer than 6 per cent of households investing remittances in business or productive assets.[30]

Conclusion: Will Climate Change Create Refugees?

Each year Bangladesh has more than a half million migrants going to work abroad and another half million moving from rural areas to the cities. Migration is the biggest cause of urban growth and especially the explosion of Dhaka as a megacity. As Box 8.1 shows, the main cause of migration is economic.[31] Most move to find a job, earn money and escape poverty. So far there are no 'climate change migrants' because climate change has not yet had a noticeable impact, and there are relatively few who migrate primarily because of environmental events such as floods, erosion, waterlogging and cyclones. However, many of them are indirectly 'environmental migrants' because the poverty they are escaping has been exacerbated by environmental events. Loss of land, livelihood or home starts a process of impoverishment that can only be solved by migration. The next two chapters look at the cities they move to – and especially the slums they live in.

Climate change with increased rain, floods and heat as well as more serious cyclones means more loss, more impoverishment and more migration. Some urban migration is inevitable. But many people do not want to move; however, they feel they have no choice. Increased protection and adaptation, and more support for those affected by disasters, would reduce the migration – but who will pay? As we have shown, Bangladesh has made remarkable efforts with cyclone shelters, flood protection and new rice varieties. But it remains a poor country – so halting future climate change migration will depend on fund-ing from those countries responsible for global warming. 'Twenty million people could be displaced [in Bangladesh] by the middle of the century,' Finance Minister Abdul Muhith said in 2009. 'We are asking all our development partners to honour the natural right of persons to migrate. We can't accommodate all these people – this is already the densest

---

29 BBS, *Survey on the Use of Remittance*, 55.
30 BBS, *Survey on the Use of Remittance*, 52, 72.
31 Bangladesh Bureau of Statistics data is consistent with other studies. See, for example, Shahadat Hossain, *Urban Poverty in Bangladesh* (New York: I. B. Tauris, 2011), chapter 6 and elsewhere.

country in the world.'[32] He called on the UN to redefine international law to give climate refugees the same protection as people fleeing political repression.

The European Union refugee crisis in 2015–16 shows that Europe is unlikely to accept climate change refugees. In any case, a more sensible alternative would be to cap global warming at levels significantly lower than those agreed in Paris in 2015, and to provide funding for adaptation and mitigation to deal with the global warming already created – to stop people needing to migrate.

32 Harriet Grant, James Randerson and John Vidal, 'UK Should Open Borders to Climate Refugees, Says Bangladeshi Minister', *Guardian*, London, 4 December 2009. Accessed 11 September 2016, https://www.theguardian.com/environment/2009/nov/30/rich-west-climate-change.

Chapter Nine

HOW CAN THE PRIVATIZED MEGACITY COPE WITH CLIMATE CHANGE?

It took only an hour and a half to go the 7 km from the hotel to Dhaka University – normal in Dhaka. But it rained during our interview. Not an unusual rain – a normal Dhaka shower. But when we left, many roads were flooded and the city was grid-locked. Most traffic is private cars, but there are many cycle rickshaws and three-wheel taxis called CNGs (after their fuel – compressed natural gas) as well as a few buses and lorries.

Legally or illegally, formally or informally, almost all spaces in this city are private. Even public space is informally privatized. We may be caught in a traffic jam but we each have our own private bit of road space, according to class. The poor sit in their rickshaws, a few mini buses and buses carry those going longer distances, office workers are in the CNGs, some ride motorcycles and the better off are in their cars. Caught in the traffic, people still work. Those in cars with drivers are on their mobile telephones; the woman in the CNG next to us is editing a report. But we are caught in a jam. At one point there were four ambulances with sirens screaming caught in the jam with us. Rich and poor, well and sick all move at the same speed.

Dhaka is the 11th largest city in the world and the densest megacity (see Box 9.1) but it is largely without mass transport. The national bus company has only 263 buses running city services in Dhaka;[1] after more than a decade of discussions, work on a first Japanese-aid-funded metro line began in 2016, as did work on a bus rapid transit line. A 2012 survey at ten major road junctions found 61 per cent of vehicles were motorized personal transport,[2] serving only the best off 7 per cent of the city's population. Yet in the first decade of the twenty-first century, the government of Bangladesh concentrated on constructing $100 mn worth of flyovers in Dhaka, hoping to improve the movement of private cars. In its greenhouse gas reduction pledge submitted before the Paris COP in 2015 (known as 'Intended Nationally Determined Contributions'), Bangladesh proposed to reduce greenhouse gas emissions by 'building of expressways to relieve congestion' as well as 'public transport measures'; Dhaka expressways and Dhaka public transport would receive equal amounts of money, $2.7 bn each.[3]

---

1   Shahin Akhter, 'BRTC Fails to Ease Public Suffering', *New Age*, Dhaka, 5 September 2015.
2   Md. Akter Mahmud, 'Mass Transit Challenges of Dhaka City', in *Bangladesh Urban Dynamics*, ed. Hossain Zillur Rahman (Dhaka: Power and Participation Research Centre, 2012). The 61 per cent motorized personal transport is composed of 31 per cent private cars, 5 per cent jeeps, 9 per cent motorcycles and 17 per cent CNGs. For the 7 per cent, Mahmud cites *Urban Transport Policy: The Strategic Transport plan for Dhaka* (Dhaka: BCL and Louis Berger Group, 2005).
3   Ministry of Environment and Forests (MOEF), 'Intended Nationally Determined Contributions', 2015, 6, 14. Accessed 14 April 2016, http://www4.unfccc.int/submissions/INDC/Published%20Documents/Bangladesh/1/INDC_2015_of_Bangladesh.pdf.

Table 9.1 Bangladesh cities – basic data

	Population, million, 2015	Annual growth rate, 1995–2015	Population density – people per km² (*)	% of Bangladesh urban population
Dhaka	17.6	3.74%	43,500	32.0%
Chittagong	4.5	2.83%	28,500	8.3%
Khulna	1.0	−0.45%	14,300	1.9%

	Urban annual growth rate 1995–2015	% of population in urban areas 2014	% of urban population in slums 2014
Developing regions	2.77%	48.4%	29.7%
Least developed countries	3.96%	36.6%	...
Southern Asia	2.67%	34.8%	31.3%
Bangladesh	3.74%	34.3%	55.1%

Note: Dhaka is the 11th largest city in the world according to UN Habitat and the 16th largest according to Demographia (*).

Dhaka has the highest population density of any of the megacities.

Most large countries have the urban population spread across several cities, but Bangladesh is very centralized. Dhaka accounts for 32 per cent of all urban population; of the megacities, only Cairo is higher, at 51 per cent.

"Developing regions" is the entire world except northern America, Europe, Japan, Australia and New Zealand. "Least developed" is a smaller group which includes Bangladesh.

Source: Tables for UN Habitat World Cities Report 2016 http://unhabitat.org/urban-knowledge/global-urban-observatory-guo/

Except: (*) *Demographia World Urban Areas: 11th Annual Edition* (Belleville, IL, US, 2015). Accessed 28 January 2016, http://www.demographia.com/db-worldua.pdf.

On the roads, there are few rules and even those seem negotiable. Traffic lights are usually ignored – if they work, which they often do not because of the inability of the city corporation to sign a maintenance contract.[1] On dual carriageways, CNGs start going the wrong way and then cars follow – only to create further confusion at the next junction. In this anarchy there are tiny islands of order and discipline: police at intersections, armed only with swagger sticks, are obeyed as they direct traffic.

It took us four hours to go 7 km back from the university; it would have been faster to walk. But that is not possible. The pavements (sidewalks) have been informally privatized and are occupied by kiosks, street traders, rag pickers, piles of building materials, rubbish and innumerable obstructions. So people join the rickshaws, CNGs, and cars and walk on the road.

But why did such a normal shower bring such chaos? Only one-third of the city is higher than 6 m above mean sea level, and, on average, Dhaka receives 2 m of rain per year, most in June, July and August. Impervious surfaces such as paved roads and roofs mean that much of the rain does not sink into the soil and runs off into lower lying areas

---

4   Shahin Akhter, 'Dhaka City Traffic Signals in Mess', *New Age*, 3 March 2013.

Box 9.1 Is Rana Plaza the true symbol of Dhaka?

The eight-storey Rana Plaza building on the edge of Dhaka represented both sides of Dhaka's story. Bangladesh's dynamic garment sector accounted for 82 per cent of the country's exports in 2014–15 and its garment factories employ 4 million people.[5] Rana Plaza was part of this sector, and its five factories employed 5,000 people. But on 24 April 2013, the building collapsed, killing 1,130 people and injuring 2,500.

Sohel Rana, the owner of the collapsed building, had been seen as a success story of the unregulated capitalism of Dhaka. Rana's father was a poor peasant who migrated from the village to Dhaka; his son had a large building which housed thousands of jobs.[6] But it became a free market, unregulated tragedy.

The *Dhaka Tribune*[7] reported that Sohel Rana was 'a member of the ruling Awami League [and] has a lot of influence'. The BBC was stronger: 'The 35-year-old Mr Rana has been described in the local media as the archetypal Bangladeshi muscleman, known locally as a "mastaan", or neighbourhood heavy. His power, influence and money came from providing muscle to local politicians. "He is known as a thug, a gangster in the area," Firoz Kabir, chairman of the Savar sub-district council, told the BBC.'[8]

He used that clout in 2007 to start construction of Rana Plaza without permission and illegally on a floodplain pond. It was originally four stories and built for commercial uses and not to take machinery, so it did not have many internal columns; the lower floors still had a shopping centre. Four floors were later added, again without permission, and using substandard materials and no internal columns. Garment factories moved to the top five floors of Rana Plaza, making clothing for retailers including Benetton, Mango, Primark, Walmart and J. C. Penny.

Dhaka is plagued by irregular electricity supplies, so the garment factories put generators on the roof. Just before 8 a.m. on the morning of 24 April there was a power cut and the diesel generators were turned on. The shaking caused cracks to appear and the building collapsed.

When Rana Plaza was built, thousands of jobs were created; the owner was an 'influential person', which meant that no one challenged that he took floodplain land that should absorb water rather than allow the city to flood, and he built an inappropriate and dangerous building. Eventually it all came tumbling down. Will this be the story of Dhaka?

---

5   Bangladesh Garment Manufacturers and Exporters Association (BGMEA). Accessed 2 February 2016, http://www.bgmea.com.bd/home/pages/tradeinformation.

6   Jim Yardley, 'The Most Hated Bangladeshi, Toppled from a Shady Empire', *New York Times*, 1 May 2013, http://www.nytimes.com/2013/05/01/world/asia/bangladesh-garment-industry-reliant-on-flimsy-oversight.html.

7   Syed Zain Al-Mahmood, 'Nexus of Politics, Corruption Doomed Rana Plaza', *Dhaka Tribune*, 26 and 27 April 2013, http://www.dhakatribune.com/politics/2013/apr/26/nexus-politics-corruption-doomed-rana-plaza.

8   Sabir Mustafa and Shyadul Islam, 'Profile: Rana Plaza Owner Mohammad Sohel Rana', BBC Bengali Service, 3 May 2013, http://www.bbc.co.uk/news/world-asia-22366454.

and then into a series of *khals* (canals) and lakes spread across the city, which then drain into the surrounding rivers. But between 1978 and 2009, about 60 per cent of wetlands and 65 per cent of rivers and khals were lost in the Dhaka Metropolitan Area; most of this occurred in the decade 1998–2009, due to unplanned urbanization.[9] The World Bank estimates that Dhaka had 43 natural canals, but 17 of these no longer exist.[10] So there is nowhere for the water to go.

This is reported; newspapers frequently contain articles about buildings being built on the flood plain or being constructed blocking drains – totally illegally. But the owners and builders are usually 'influential people' and the inspectors and planners have been paid to turn a blind eye to the illegal constructions. Every piece of land is grabbed for building – high-rise blocks of flats in better-off areas, multistorey factories and vast, densely packed slums with precarious three-storey buildings made of bamboo. Just as on the road, rules relating to drainage, water, electricity, zoning, construction quality and pollution are largely ignored.

And Dhaka is a magnet. Unusually for large countries, Bangladesh has a single large and dominant city, which concentrates commerce, industry and government. There is only one other city, Chittagong, with over 1 million people. Dhaka attracts people because they can find work – women in the garment factories and men as rickshaw pullers. Two-thirds of Bangladesh's urban manufacturing employment is in the Dhaka metropolitan area.[11]

Bangladesh is still a rural country, but the rural population has been steady at 105 million for more than a decade. All the population growth has been in urban areas, which have been growing at 3.5 to 4.5 per cent per year for three decades (see Figure 9.1).

The Politics That Create the Least 'Liveable' City

The Economist Intelligence Unit rates Dhaka as one of the least 'liveable' cities in the world – out of 140 cities, only Lagos and war-shattered Tripoli and Damascus are worse.[12] 'Rapid mass urbanization is occurring without development,' argues Shahadat Hossain in his book *Urban Poverty in Bangladesh*.[13] This means that all of the environmental management successes reported so far in this book – such as cyclone shelters, rice

9 Mallik Sezan Mahmud et al., 'Remote Sensing & GIS Based Spatio-Temporal Change Analysis of Wetland in Dhaka City, Bangladesh', *Journal of Water Resource and Protection*, 3 (2011): 781–87, doi:10.4236/jwarp.2011.311088.

10 Sarwar Jahan, '*Climate and Disaster Resilience of Greater Dhaka Area: A Micro Level Analysis*', Bangladesh Development Series Paper No. 32 (Dhaka: World Bank, 2015): 1.

11 M. A. Taslim and Akib Khan, 'Industrial Growth and Location Dynamics', in *Bangladesh Urban Dynamics*, Rahman ed., 65.

12 Economist Intelligence Unit, *A Summary of the Liveability Ranking and Overview: August 2016* (London: Economist Intelligence Unit).

13 Shahadat Hossain, *Urban Poverty in Bangladesh* (London: I. B. Tauris, 2011), 191.

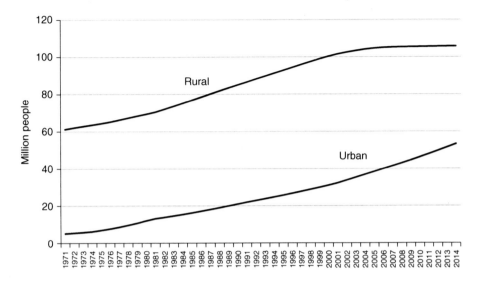

Figure 9.1 Bangladesh rural and urban population
Source: World Bank data bank.

production and tidal river management – are rural. It is in the rural areas and smaller cities with a million or fewer people that the adaptation to environment and climate change is occurring, while the two big cities, Dhaka and Chittagong, are moving backward environmentally and are less able to cope with climate change. How did this happen? This is subject to extensive debate and many reasons are put forward. We argue that Bangladesh's politics and political history has led to economic growth and real gains in rural areas, but has proved ill-adapted to urbanization.

What is now Bangladesh had been marginalized by the British colonial state and then by West Pakistan; it was then decimated by the massacres of the local elite in 1971, so a high priority of the independence government was to build a domestic bourgeoisie. The mood of the 1950s and 1960s, both in Europe and in thinking about development in poorer countries, was social democratic, with governments intervening to provide services, promote growth and reduce inequality. The first elected prime minister, Mujibur Rahman, pushed for a socialist state and was allied to India and the Soviet Union. In 1975, he was overthrown and killed by the military, which ruled until 1990. The military adopted more free market and pro-Western policies. In 1991 the Bangladesh Nationalist Party led by Khaleda Zia won the election, and in 1996 the Awami League, headed by Sheikh Hasina, one of Mujib's surviving daughters, was victorious. Both governments further liberalized and also responded to demands from Western aid donors to adopt the neoliberal model, a pressure that was particularly strong at the end of the cold war. The World Bank and IMF called for a lesser role for the government, for deregulation to set free the private sector, and for growth driven by increasing the wealth and power of the elite. A clientelistic political system was inherited from Pakistan, expanded and developed by the military, then built on by elected governments and continues until today. In the local interpretation of the neoliberal model, state resources such as land, contracts and jobs were assumed to be the route to build a new domestic capitalist class – linked to whichever party was in power. Civil servants, seeing what the elected elite were doing, also aligned themselves with whichever party was in government and began to take a share of public resources.

But Mirza Hassan of the Institute of Governance and Development at BRAC University points to the supposed 'paradox' that a weak, inefficient, corrupt state with bad governance has increased its per capita income fourfold and cut poverty by more than half[14] (see Table 1.1). Transparency International ranked Bangladesh number one for corruption in five consecutive years (2001–6) but economic growth increased from 4.8 per cent to 6.5 per cent. Hassan points to three 'elite political settlements' that have shaped the past 25 years of Bangladesh development. The first is that 'Bangladesh exhibits a remarkable and long-standing elite consensus on pro-poor development strategies that is manifested not only in political rhetoric, but also in actual budget allocations.' This is partly political because it allows the doling out of patronage and gains political support, building up a voting bloc or 'vote bank' that will support the party in the next election. And it is linked to memories of the 1974 famine and the need to ensure food security and implement rapid social protection measures as a way of ensuring legitimacy. But patrons also have social, customary and moral obligations to their clients, and local elites in rural areas have direct contact with the recipients of assistance. Thus this combined political and social network promotes development in rural areas, but, as we note below, this does not seem to work in urban areas.

Hassan's second elite political settlement is that 'nurturing and maintaining a robust private sector has been one of the top priorities of all political regimes'.[15] At the same time, business people have integrated into politics and the state. In the 2009–13 government, at least one-tenth of MPs were garment factory owners.

The third settlement is to accept a 'partyarchy' in which the two parties are antagonistic and compete, but civil society and state institutions are politicized along party lines. There is a duopoly of power and the two sides alternate. At all levels, organizations respond to this; businesses, civil society and trade union leaders – down to informal local leaders – either change party or accept that they are replaced by someone allied to the other party or allied to a new minister or MP. Hassan notes that 'once in power, leaders tend to capture most of the state, educational, cultural, social and economic institutions and indulge in large-scale patronage distribution among party members and supporters.' This is an essential part of the patron-client system. Businesses try to co-opt political actors from both parties as directors, and business association leaders are also closely linked to the governing party. Many businesses that are dependent on government contracts engage political negotiators to organize contracts, informal payments and subcontracts that need to go to firms owned by ruling-party leaders.

---

14 Mirza Hassan, 'Political Settlement Dynamics in a Limited-Access Order: The Case of Bangladesh', ESID Working Paper No. 23 (Manchester, UK: Effective States and Inclusive Development Research Centre (ESID), School of Environment and Development, University of Manchester, 2013).
15 Ibid.

Box 9.2 Rivers around Dhaka in death throes

'Turag River – one of the four rivers surrounding the capital city Dhaka – has been reduced to a stagnant cesspit as encroachers and factories continue to encroach upon and pollute the river,' wrote the Dhaka daily, *New Age*, as part of a campaign in early 2016.[16] There has been so much construction along the river that it is no longer navigable. 'During a recent visit […] factories located on both sides of the river were discharging toxic liquid waste directly into the river.' The journalists found the river was effectively dead, with a dissolved oxygen concentration of 0; the minimum for fish to survive is 5 mg per litre. The Sitalakhya River on the east side of Dhaka is also dead; it has hundreds of industries, including fertilizer, textile and dyeing factories along its banks, dumping waste into the river. Many of the factories were built on illegally filled riverbed land.

Taqsem A. Khan, managing director of the WASA agrees there is a problem.[17] Of Dhaka's drinking water, 87 per cent comes from ground water wells, but the aquifer is being depleted and the layer of ground water is falling by 2–3 m per year. This is because urbanization means there is much less open land so less water soaks in to recharge the aquifer. He would like to take more water from the rivers, but it is too polluted 'due to indiscriminate discharge of domestic waste water and industrial effluent'. So the water intake will have to be 60 km away from the city. Khan admits WASA is also part of the problem. Only 30 per cent of households are connected to sewerage pipes that lead to the single large treatment plant. But the draft sewerage master plan proposed in 2014 largely ignored the slums, which are a main source of untreated sewage. WASA faces one other problem – irregular electricity to run its pumps – so it had to buy 200 generators.

In 2009 the High Court directed the government to take appropriate steps to stop encroachment, earth-filling and illegal building on the rivers surrounding Dhaka, but *New Age* reports that government is actually supporting encroachment. A government review committee had identified flood zones and water bodies not to be built on, but in 2015 a cabinet committee approved plans by a private company to build two large new towns on the flood basin area. Everyone points the finger at someone else: Dhaka Deputy Commissioner Tofazzal Hossain Mia told *New Age* that the city had done nothing to recover illegally occupied land along the Balu River because the Bangladesh Inland Water Transport Authority (BIWTA) had not asked it to. *New Age* notes that land filling has taken place next to the BIWTA headquarters on the Turag River.

---

16 Mahamudal Hasan and Shahin Akhter, 'Rivers around Dhaka in Death Throes,' 27 January 2016; 'Encroachment, Pollution Leave Turag Tottering', 28 January 2016; and 'Govt Among many Actors Destroying Balu River', 29 January 2016, *New Age*.

17 Taqsem A. Khan, 'Dhaka Water Supply and Sewerage Authority: Performance and Challenges', Dhaka WASA, 2012. Accessed 10 September 2016, http://dwasa.org.bd/wp-content/uploads/2015/10/Dhaka-WASA-Article-for-BOOK.pdf.

The result was a dynamic and growing economy, particularly driven by the garment sector. But as Box 9.1 shows, huge problems have been created by an informalized free-market Dhaka.

Civil society and nongovernment organizations (NGOs) also have a role in the partyarchy and the neoliberal approach. Social services are provided by giant NGOs that have become similar to international corporations. The incentive is for 'NGO officials to remain loyal to the existing elite political settlement, rather than voice protests against it', comments Hassan.[18] In rural areas NGOs support the top-down patron-client system, and try to work with the grain of rural power structures to gain benefits for poor people – especially their own members. They rarely rock the boat by pushing for rights-based approaches and claims of the poor; they do not promote the mobilization and organization of the poor and have become a social rather than a political force.

Shapan Adnan, in a study of land-grabbing by the elite for shrimp farming in Noakhali, reported that 'members of these land-grabbing interest groups were affiliated to not only the current ruling party, but also the major opposition parties. This is indicative of *collusion* and division of the spoils among all the major political fractions within these classes, rather than discriminatory patronage channelled to the followers of any particular political party or fractional grouping'.[19] Thus the partyarchy settlement assumes a certain shared interest among the elite. Both sides accept that governments will change hands, and that each group will eventually gain patronage power, but that the opposition is never totally excluded. Thus they are prepared to allow major long-term projects agreed by the previous government to go ahead (perhaps with changed or additional bribes.) In 2014 the main opposition party, the BNP (Bangladesh Nationalist Party), boycotted the general elections and Sheikh Hasina's Awami League was re-elected effectively unopposed, but it is not clear whether this has disrupted the partyarchy settlement.

Politics and patronage in Bangladesh is tightly controlled from the top, which has led to the development of Dhaka as what is called a 'primate city', which is the largest city in its country or region and which is disproportionately larger than any others. Without government planning and spatial incentives, businesses tend to congregate in the biggest centres with the largest supply systems and markets. Dhaka is also the centre of government and politics; 'Dhaka's inexorable growth as a primate city is mirrored in the extreme centralization of decision-making and political authority,' notes Hossain Zillur Rahma.[20] He was a minister in the 2008 caretaker government (see Chapter 2) and set up the Bangladesh Urban Institute in 2012.

---

18 Ibid.

19 Shepan Adnan, 'Land Grabs and Primitive Accumulation in Deltaic Bangladesh: Interactions between Neoliberal Globalization, State Interventions, Power Relations, and Peasant Resistance', *Journal of Peasant Studies* 40 (2013): 107.

20 Hossain Zillur Rahman, 'Urbanization in Bangladesh: Challenges and Priorities', paper presented at the Bangladesh Economists' Forum, Dhaka, 21–22 June 2014. Accessed 13 September 2016, http://www.pri-bd.org/upload/file/bef_paper/1414213688.pdf.

Finally, despite having a global megacity where the Bangladesh elite make their fortunes, it has not really accepted urbanization. It is still a majority rural country, and as Hossain Zillur Rahman explains in his book *Bangladesh Urban Dynamics*, 'many of our policy-makers continue to interpret the reality of urbanization as an unwelcome story to be resisted rather than managed as a driver of change for inclusive growth and sustainable environments. Consequently, a peculiar political economy dominates, characterized by a policy ambivalence on one hand and the entrenchment of a power nexus on the other that makes unplanned growth and poor urban governance the norm rather than the exception.'[21] 'The government prioritizes the richest class in the city,' argues Dibalok Singh, executive director of the NGO Dushtha Shasthya Kendra (DSK). 'They do not like the slums, and they think that to provide basic services will encourage a migration to the city, and that they can halt migration by not providing services,' he told us. For example, 24.6 per cent of extremely poor Bangladeshis are covered by the social safety net, but in urban areas it is only 9.4 per cent.[22] The elite still sees rural areas as the rightful home of the poor, and this view is reinforced by an image of crime and squalor, rather than of working people making the best lives for themselves and their children.[23]

Shortly after taking office in 2009, Finance Minister Abul Maal Abdul Muhith told a conference that previous governments were concerned with rural poverty, 'but now it is needed to concentrate on urban poverty as urban areas are going to be worst hit due to indirect impact of climate change'.[24] But this did not happen. Indeed, cities are sometimes completely ignored. The 2015 World Bank-funded Regional Weather and Climate Services Project[25] involves the Bangladesh Meteorological Department, the Bangladesh Water Development Board, and the Department of Agricultural Extension (DAE) but does not deal with cities at all. The rural-urban split is clear in the *Bangladesh Climate Change Strategy and Action Plan 2009*,[26] which has a specific project to help 'poor and marginal farmers' but nothing similar for the urban poor.

This section started with the point that Dhaka is the least liveable megacity, and that Bangladesh is a paradox because of growth and poverty-reduction despite bad

---

21 Hossain Zillur Rahman, 'Bangladesh Urban Dynamics: Exploring a Holistic Perspective', in *Bangladesh Urban Dynamics*, ed. Hossain Zillur Rahman (Dhaka: Power and Participation Research Centre, 2012), 1.

22 Syed Hashemi, 'Can Asset Transfer Promote Adaptation amongst The Extreme Urban Poor? Lessons from the DSK-Shiree Programme in Dhaka, Bangladesh', in *Urban Poverty and Climate Change*, ed. Manoj Roy et al., (Abingdon, UK: Routledge, 2016), 218.

23 Nicola Banks, Manoj Roy and David Hulme, 'Neglecting the Urban Poor in Bangladesh: Research, Policy and Action in the Context of Climate Change', *Environment and Urbanization* 23 (2011): 499.

24 'Urban Areas to Be Worst Hit by Climate Change: Muhith', *Daily Star*, 29 January 2009.

25 Marie Florence Elvie, 'Bangladesh – Regional Weather and Climate Services Project: Environmental Management Framework'. Bangladesh, 2015 (no publication details, World Bank Project P150220). Accessed 12 March 2016, http://documents.worldbank.org/curated/en/2015/12/25744289/bangladesh-regional-weather-climate-services-project-environmental-management-framework.

26 Ministry of Environment and Forests, *Bangladesh Climate Change Strategy and Action Plan 2009* (Dhaka: Ministry of Environment and Forests, 2009), Programme P1T9.

governance. The preceding political discussion shows that while the support for businesses is urban, the support for pro-poor development strategies is largely rural, and
the elite sees 'the people' as being largely rural. Partyarchy and its patronage, thuggery
and corruption is moderated in rural areas where there are stronger social controls,
and where elected representatives to upazilas and union parishads (UPs) live locally
and have local friends and family. This serves as a check on egregious behaviour. Each
UP councillor represents a ward with only 3000–4000 people.[27] In a patronage society,
the councillors' social standing depends on doing the right things for their constituents
and ensuring that corruption and pressure from thugs is not too harmful. Actions of
politicians, thugs and corrupt business people are tempered. As noted in Chapter 4, in
water disputes, local people go first to elected bodies, and not the special water committees imposed by donors that are seen as having been captured by elites. The move to a
grassroots-driven tidal river management has succeeded because local politicians backed
community organizations. But in urban areas political processes are very different.

Smaller Cities

In smaller cities, some of this direct social and political contact still works, but it is more
based on building vote banks – blocs of loyal voters. Rajshahi is a small city of 450,000
people in the west of Bangladesh.[28] The Rajshahi City Corporation is divided into
30 wards, each with a councillor, plus 10 elected women (one for every three wards).
Thus each councillor represents 15,000 people, rather more than in a UP, but a lot less
than in Dhaka. Parvaz Azharul Huq of Rajshahi University studied the politics of the
city.[29] In two of four communities studied, the councillor visited regularly and talked to
constituents. More than half of Rajshahi residents live in slums (see Chapter 10)[30] and
eviction is a constant threat. In one slum area, the councillor actively supported the
people 'and gave them protection while eviction threats came up. In return, the settlers
basically acted as followers and a power base for the WC [ward councillor].'

27 There are 4,544 union parishads with an average 2010 population of 27,463. Each is divided
    into nine wards, each of which elects one councillor; in addition three women councillors
    are elected. Mohammed Mamun Rashid, 'Union Parishad (UP) Election and Communities
    View: A Micro-Level Study in Rural Bangladesh', *Crown Research in Education*, 2 (2012): 158–64
    and Commonwealth Local Government Forum, 'Bangladesh', London, 2013, http://www.
    clgf.org.uk/userfiles/1/file/Bangladesh_Local_Government_Profile_2013_CLGF.pdf.
28 Bangladesh Bureau of Statistics, *Bangladesh Population and Housing Census 2011*, National report,
    Volume – 1 Analytical report (Dhaka: Bangladesh Bureau of Statistics, Ministry of Planning,
    2015). 'City Corporation Area'. Defining the size of a city depends on definition. The figures
    in Table 9.1 refer to broader urban areas, which often include neighbouring towns that are
    part of the metropolitan area of the city.
29 Parvaz Azharul Huq, 'The Limits of Citizen Participation in the Urban Local Governance
    Process in Bangladesh', *International Journal of Public Administration*, 37 (2014): 424–35.
30 Bangladesh Bureau of Statistics, *Bangladesh Population and Housing Census 2011*, National report,
    Volume – 1 Analytical report, table H03. Housing type is not available for city corporations.
    Using the somewhat larger urban populations of Rajshani and Khulna divisions, housing by
    share of families was: Rajshani: pucca, 23 per cent, semi-pucca, 37 per cent, kutcha and jhupiri,

Khulna in the south, Bangladesh's third largest city but with a 2011 population of only 660,000, shows similar features. In the Rupshaghat slum that grew behind a new embankment, ward councillors have pushed for municipal services and protected residents against eviction. But it has sometimes required large demonstrations in front of the press club or outside the residences or offices of politicians.[31] Lower land prices in smaller cities also make it possible for smaller landlords, sometimes living on the site, to develop small rental complexes – some of good quality and some poorly constructed with inadequate water and sanitation. Residents can stay so long as they are on good terms with the landlord and pay the rent, but with most people doing day labour or informal work, this is not always possible. A survey in the Bagmara private settlement of 70 households in Khulna showed that the average tenancy was only two years. Bagmara residents failed to make links with influential political and civil society actors and did not register to vote, so they were powerless.[32]

Even in the smaller cities, surveys show that eviction is still the main fear of slum residents. A survey in Rupshaghat showed eviction threats as the biggest problem, drug selling came next, and only then water and sanitation.[33] Climate change does not even register as an issue, because there are much more pressing problems. Nevertheless, communities in smaller cities do appear to have some influence over their environment.

But small cities are not growing, because people want to go to Dhaka or Chittagong, despite government and donor attempts to support the small cities. Hossain Zillur Rahman points to the huge income gap between big and small cities. The average income in secondary towns is just 10 per cent higher than in rural villages, while the average income in Dhaka and Chittagong is 69 per cent higher than in secondary cities and 87 per cent higher than in villages.[34] Rajshahi grew by 15 per cent between 2001 and 2011, while Khulna actually declined in size by 9 per cent.[35]

Dhaka and the Mastaans

Even if they will be powerless, the migrants flow toward the megacities,. Each Dhaka ward councillor represents about 120,000 people.[36] Politicians do not live locally and their social networks are less likely to impinge on those affected by their policies. Elected

---

40 per cent; Khulna: pucka, 32 per cent, semi-pucka, 34 per cent, kutcha and jhupiri, 34 per cent. See Chapter 10 for definitions; slums would be considered to be all of kutcha and jhupiri and some of semi-pucka.

31 Manoj Roy, David Hulme and Ferdous Jahan, 'Contrasting Adaptation Responses by Squatters and Low Income Tenants in Khulna, Bangladesh', *Environment and Urbanization* 25 (2013): 167.

32 Manoj Roy, Ferdous Jahan and David Hulme, 'Community and Institutional Responses to the Challenges Facing Poor Urban People in Khulna, Bangladesh in an Era of Climate Change', BWPI Working Paper 163 (Manchester, UK: Brooks World Poverty Institute, University of Manchester, 2012): 9, 17.

33 Banks, Roy and Hulme 'Neglecting' 496.

34 Rahman, 'Bangladesh Urban', 5.

35 Khula and City Corporation Areas, Bangladesh Bureau of Statistics, *Bangladesh Population and Housing Census 2011*, National report, Volume – 1 Analytical report, table 3.4.11, 45–52.

36 Dhaka is divided into two halves, north and south. In the 2015 elections Dhaka South had 57 wards and a population of about 7 million. There is one councillor for each ward, plus one woman councillor for each three wards.

officials work through intermediaries, known as *mastaans*. The literal translation of mastaan is 'muscleman' or 'enforcer', and they serve both as intermediaries and as thugs who use violence to enforce the will of politicians and influential people. To gain access to slums, NGOs often have to negotiate with the mastaans.

Shahadat Hossain studied an unidentified Dhaka slum in detail and showed the complexities and permanent conflict over informal control.[37] Local musclemen took over 23 hectares of vacant state land in the 1990s and leased it out. There were 15,000 families and a market of 400 shops. In this slum, Dhaka Water Supply and Sewerage Authority (WASA) and Dhaka Electricity Supply Company (DESCO) did not provide any direct utility supplies, so the musclemen also installed an illegal water connection and water tank, and an illegal electricity connection. The mastaans controlled the shops and kept increasing the rents; they also became involved in illegal activity including drugs and rape. In 2003, the shopkeepers organized a response; two of the mastaans were killed and a shopkeepers committee took over the slum. This committee was in turn taken over by local influential people linked to the BNP party, which was in government at the time, and when the Awami League (AL) won the election at the end of 2008, it was taken over by influential people belonging to AL. The water tank and supply changed hands several times, eventually being taken over by three influential people who rented out the tank to local families whose children sold water all day from the tank. There are a few piped connections, mainly to homes of influential people who receive free water, but there has been no improvement in water supply to ordinary residents. A semi-legal electricity connection was expanded and reorganized to distribute electricity to 43 local leaders, who in turn sold it at higher prices to shops and shanties – the price to shop owners was 6 taka ($.08) per lamp or fan per day.

Hossain continues: 'Inhabitants' dependency on local leaders is not limited to getting access to utilities and business in the bazaar, but rather it importantly includes favours to mitigate the effects of a community level conflict, contacts with the police station in case a family member gets arrested, preferential treatment in relief programmes implemented by government departments (usually performed through the political party in power), requests to local government leaders for elderly allowance, support for the admission of a child to NGO run local schools, etc.'

The lives of the poor living in slums are controlled by a series of local power brokers. Eviction is the biggest fear, but these power brokers also control access to water and electricity, and to commercial spaces that many need to earn their meagre living. Rickshaw pullers, construction workers and other day labourers often work through labour brokers – they take their commission, but both sides find them useful because the labourer does not have to search for a job each day, while the rickshaw owners know the broker will ensure the rickshaw is not stolen or damaged. Power brokers often have political links and put pressure on their clients to attend rallies and vote for certain candidates. In these patron-client relations, the poor gain some small support in event of illnesses and win limited protection against eviction. Parties build up vote banks of urban low-income residents largely from negative aspects, such as protection from eviction and the police, whereas in rural areas it comes more from the implementation of pro-poor policies.

---

37 Shahadat Hossain, 'The Production of Space in the Negotiation of Water and Electricity Supply in a Bosti of Dhaka,' *Habitat International* 36 (2012): 68–77.

The urban elite likes the status quo and its class divisions, Binayak Sen, research director of the Bangladesh Institute of Development Studies (BIDS), explained to us. In the present system, 'each class has its own transport and there is no mixing. Metros, as in European cities, are more democratic and everyone uses them. So here, the elite is not interested in inclusive public transport'. Indeed, it is a vicious cycle. There are cuts to the water supply, so the better off buy pumps, which require electricity which also suffers cuts, so they buy generators. They live in nice flats on upper floors, so are less affected by floods. And, of course, they have internet and mobile telephones. Generators, pumps and big cars are essential and allow the better off to largely ignore the urban crisis, but these cost money, so the rich need more money to fund their segregation. But when they are no longer dependent on these services, they apply no pressure to make them work.

Nevertheless, everyone is caught in the floods and traffic jams, which means the rich are fouling their own nests. Prof. Anu Muhammad, an economist at Jahangirnagar University, took a hard line when he told us that 'elites do not see their future here in Bangladesh – the next generation will live abroad so they don't care.' They send their children to universities abroad and assume they will not return to Bangladesh; many have houses abroad, at least in Malaysia. 'They don't care about the consequences of land grabbing and blocking canals, floodplains and parks.'

Rivers and Floods

Gulshan Lake kept rising, and soon adjoining streets of exclusive residential areas were knee-deep in water. People could not use their cars. Diplomats and wealthy people had to move about in rickshaws or high vans. Flooding started on 22 July 1998 and continued for 65 days; the highest level was reached on 12 September. Even in the exclusive areas of Gulshan and Banani, flooding continued for a month. Some people hired boats. Traffic jams were replaced by boat jams as 4 million people became dependent on boats instead of rickshaws. In the eastern part of Dhaka, which was totally flooded, boats became the main means of transport. There were health problems as sewerage became mixed with flood waters. The 1998 flood was the worst in a century not just for Dhaka, but for the whole of Bangladesh (see Figure 1.1). An excellent book, *The 1998 Flood*, serves as the basis for this section.[38]

Flood-affected middle-class people largely moved out to live with relatives; poor people built temporary platforms inside their houses or moved up to the roof if the house was strong enough, and when the water rose too high they moved to the higher ground occupied by the better off (who allowed them to stay) or moved to temporary flood shelters and relief camps. Well-constructed (*pucka*) houses suffered only limited damage, but houses of poorer people suffered structural damages and shacks were often destroyed.

Despite the seriousness of the flood, a survey found 'that the basic needs of the flood-affected people were largely met and they were saved from starvation and disease'.[39] National and Dhaka city governments, local councillors, NGOs, private sector shelter

---

38 Ainun Nishat et al., *The 1998 Flood: Impact on Environment of Dhaka City* (Dhaka: Department of Environment, Ministry of Environment and Forest, 2000).

39 Ibid., 225.

operators and neighbours all played a role. Authors of *The 1998 Flood* had high praise for the relief effort, but did raise three criticisms:

- 'Political leaders and workers, specially those from the ruling party [Awami League], have been found to intervene in relief distribution, where party consideration became prominent over the real need of the victims. This had resulted in improper relief distribution at many places by way of negligence towards the actual distressed and disadvantaged groups'
- NGOs insisted in doing their own distributions and did not participate in government coordination. 'This usually caused improper identification of real disadvantaged groups.' Some relief items were of poor quality.
- 'Able-bodied men jumped ahead of the queue while the weaker section stood at the end. [...] Weaker groups [...] received less or no aid in some cases and became the real victims of the flood.'[40]

A survey of slum families in 2009[41] (after the 2004 and 2007 floods) found that 'mutual help and support are dominant features in times of crises. While poorly educated and resourced slum dwellers are highly vulnerable to external shocks, they still show a surprising capacity to cope with natural calamities. [...] The very positive responses to questions regarding mutual support networks suggest that trust, reliability and reciprocity are essentials of social life in Dhaka's slums.'

A Flood-Prone Megacity

Dhaka is 200 km from the sea and is surrounded by four rivers and a *khal* (canal). These receive water from the Ganges and Brahmaputra rivers. At high tides, water from the Meghna river is also pushed back towards Dhaka. Only one-third of the city is higher than 6 m above mean sea level, and this is the most urbanized part with the permanent buildings. East Dhaka, where the main urban expansion is taking place now, and many of the slum areas are less than 6 m above mean sea level.

Dhaka's first flood protection embankment was built along the Buriganga River in 1864. But Dhaka remained largely unprotected at the time of the 1988 flood, which was the worst in a century (until it was surpassed a decade later); this triggered the Greater Dhaka Flood Protection Project. As discussed in Chapter 4, it is normal for low-lying areas like Dhaka to flood, and these areas store the flood water, releasing it slowly. But it was decided that this was not practical for urban areas, and that barriers should be built to protect the capital, in a Dutch-style system of dykes protecting polders containing the city. The western half of Dhaka, the most urbanized part of the city, was encircled with 30 km of embankments and 37 km of raised roads and flood walls. Regulators and pumping stations were also built.

---

40 Ibid., 227.
41 Boris Braun and Tibor Aßheuer, 'Floods in Megacity Environments: Vulnerability and Coping Strategies of Slum Dwellers in Dhaka/Bangladesh', *Natural Hazards* 58 (2011): 771–87.

The less populated eastern half of Dhaka was to be left to absorb the flood waters and the water was to be pumped out of the western half of Dhaka. But Dhaka is a rapidly expanding megacity, and the best available land, both for exclusive residential districts and for new slums, is in the east. This remains an issue.

The flood protection walls have clearly made a difference, protecting much of west Dhaka in the major floods of 1998 (which was worse than 1988) and in less serious floods in 2004 and 2007. In 1988, 75 per cent of West Dhaka had been flooded, although that flood lasted only three weeks. A decade later, 100 per cent of east Dhaka and only 23 per cent of West Dhaka was flooded.

The 1998 flood showed a number of problems in the new system. First, within the city the drains were not adequate to carry the huge volume of water required, nor was there sufficient pumping capacity to lift the water out when the surrounding rivers were higher than the land inside the city. Filling of drains, khals and lakes with rubbish and for new construction had exacerbated the problem. Management was also a problem, causing two areas of west Dhaka to flood; the new raised road has a series of drainage structures designed to be closed in event of high water, but they were not. Some were eventually closed, but one remained open during the entire period of the 1998 flood. In old Dhaka the dykes were not high enough, and local people erected a 1.5 m high sandbag barrier, which saved the historic city from flooding.[42]

Floods in 2004 showed a different problem. The dykes and embankments largely protected West Dhaka from river flooding. However, intense rainfall in September 2004, including 341 mm in just 24 hours, combined with insufficient pumping capacity, led to localized flooding and waterlogging inside the embankments.[43] Some high-income areas such as Gulshan and Banani, experienced prolonged inundation. The sewerage system broke down, which resulted in contaminated drinking water, posing a threat to public health. In July 2009, more than 290 mm of rain fell within six hours, setting a 60-year record and flooding major roads. One of the worst affected areas during the extreme rainfall events of 2004 and 2009 was Old Dhaka. And in all of these, an increasingly populated East Dhaka was flooded.[44]

Southeast of Dhaka is the Dhaka-Narayanganj-Demra area which was developed by Bangladesh Water Development Board (BWDB) as an agricultural irrigation project, with polders protected from floods by embankments and drained by pumps. Despite a ban on development, it is now largely urbanized, and buildings have been built without permission and do not follow building regulations. Though the area is free from external flooding and there is some pumping, waterlogging occurs during the rainy season, sometimes for long periods.

Climate change will bring sea level rise, which will have an impact on Dhaka because the rivers will not drain as fast and the river levels will be higher – meaning higher dykes

42 Nishat, *The 1998 Flood*, 27.
43 Ashraf Dewan, *Floods in a Megacity: Geospatial Techniques in Assessing Hazards, Risk and Vulnerability* (Dordrecht: Springer, 2013), ch. 3.
44 Susmita Dasgupta et al., *Urban Flooding of Greater Dhaka in a Changing Climate: Building Local Resilience to Disaster Risk* (Washington: World Bank, 2015), 4–11, 215, http://elibrary.world-bank.org/doi/book/10.1596/978-1-4648-0710-7.

will be needed. Climate change will bring more intense rainfall, so more record-breaking rainstorms are to be expected. More intense rainfall and higher rivers means more water-logging and more pumps needed.

Conclusion: Can Adequate Action Be Taken?

In a 2015 report, the World Bank says that 'since 2004, urban flooding conditions have improved in parts of Central Dhaka due to key investments. WASA has worked hard to lay new drainage pipes, reclaim encroached khals, and improve conveyance in pipes and khals through desilting.'[15] But it goes on to note that all planned improvements will only be able to cope with medium-sized rainfall events of 200–250 mm per day. In the period 2003–9, this was only exceeded twice – in September 2004 (341 mm in one day) and July 2009 (290 mm in six hours).[16] But such intense rain will likely become much more frequent with climate change. Furthermore, WASA plans hardly consider East Dhaka or the Dhaka-Narayanganj-Demra area.

The Bank calls for new pumping stations and drains, but it stresses the need to protect all water bodies and khals from encroachment and solid waste. 'To deter encroachment, the water bodies should be lined with pedestrian walkways. In addition, Dhaka City Corporations (DCCs) should be encouraged to transform walkway-lined water bodies into urgently needed, passive recreation zones, where city residents can enjoy open-air activities.' It stresses the need for better solid and human waste management.

All lovely ideas, and it would be nice to be able to walk around the many water bodies, as thousands still do every day along Dhanmondi Lake. But is there the political will to create public spaces and provide public services? And what about the poor majority who live in slums? Can Dhaka undergo the attitudinal change of Victorian London, which came about when the elite realized that everyone benefits if the poor have adequate housing and basic public services? Or is it just too profitable for influential people to assume that the poor can be sent back to the countryside and multistorey blocks of flats built where they used to live?

Dhaka is implementing one major water and public space project. The Hatirjheel is the largest water body inside Dhaka. It is located near the centre of the city and is a main holding area for monsoon rainwater. The Begunbari Khal (canal) links the lake to the Balu River. Gulshan and Banani lakes drain into the Hatirjheel. The lakes and surrounding land are below the flood levels of the surrounding rivers, so the Rampura sluice gate on the Begunbari Khal – built after the 1988 flood – is closed for two months each year during the monsoon. Thus monsoon rains from a large part of the city are kept in the Hatirjheel, and its level rises.[17] If the level rises 2 m, the lake can hold an extra 1.5 mn m³ of water from the rains.[18] The lake and canal were being lost like many others, but a

---

45   Ibid., 199–203.
46   Ibid., 215.
47   'Hatirjheel Losing Attraction', *Daily Observer*, 12 December 2014.
48   Md. Sabbir Mostafa Khan and Md. Feroz Islam, 'Potential Application of Early Warning System for Urban Flooding: Case Study of Central Part of Dhaka City', *International Journal of Engineering Research & Technology* 3 (2014): 1191.

$350 million project from 2009 to 2013 reversed that and has saved a key water storage area and created a public open space. Most illegal occupants were removed. The lake and canal were dredged and deepened and a neighbouring wetland was preserved for rain water storage. Roads and bridges were built, and along the lake, parkland was created with 19 km of footpaths and walkways. This immediately improved the monsoon water storage capacity – and it could be increased. There are already a few pumping stations to lift water out when rivers are high, and more are being built. The World Bank estimates that if forecasts were used and pumping was started before predicted heavy rain, lowering the level of Hatirjheel and linked lakes by 1 m would mean they could absorb an additional 80 mm of rainfall.[49]

Unfortunately, the Hatirjheel project has not been without problems. The biggest illegal building is the 15-storey headquarters of the Bangladesh Garment Manufacturers and Exporters Association, which was built in 1988 after an island, which occupies a large part of a key channel, was constructed. In 2013 the High Court ordered the building to be demolished, but this has not happened. The other problem is that the lake water is black and smells foul, making the lakeside a less pleasant walk than was planned. Many sewerage pipes have been illegally connected to the storm drains that flow into Hatirjheel and a planned waste treatment plant was not built; solid waste is also dumped in the lake. A study showed 'that water of the Hatirjheel lake was severely polluted', with high levels of dissolved solids and low levels of dissolved oxygen.[50] Hatirjheel is part of the boundary between the developed west side of Dhaka, which is 75 per cent built up, and the east side which is only 30 per cent built up and has 66 per cent open lands and water bodies.[51] The east is the main area of expansion, and rapid urbanization is predicted, so maintaining the large area of water and wetland is an important change of direction.

Dhaka has made some changes, but they are not enough. Profit for the rich and survival for the poor are the guidelines. It is still a city without urban pride and without a vision of how to become a habitable city. It is not yet planning for its current environmental and climate problems, and has not even begun to think about climate change. How many more floods and how much traffic grid-lock will be required to force the Dhaka leadership to think about how to deal with an exploding megacity just a few metres above sea level?

---

49 Dasgupta, *Urban Flooding of Greater Dhaka*, 203.
50 M. S. Islam et al., 'Investigation of Water Quality Parameters from Ramna, Crescent and Hatirjheel Lakes in Dhaka City', *Journal of Environmental Science and Natural Resources* 8 (2015): 1–5.
51 Sadia Afrin, Md. Maksimul Islam and Md. Mujibur Rahman, 'Assessment of Future Flow of Hatirjheel-Begunbari Drainage System due to Climate Change', *Journal of Modern Science and Technology* 3 (2015), 102–16.

Chapter Ten

IS CLIMATE CHANGE ONLY A PROBLEM FOR THE URBAN POOR?

Police in riot gear accompanied the bulldozers as they knocked down the houses in Pora Basti section of the Kallyanpur slum, Dhaka, on 21 January 2016. By mid-day, a High Court judge ordered the demolition to stop, saying that a ten-year old High Court order against demolition was still in force. Local people reported that the next day, after most residents – mainly rickshaw pullers, domestic help, street sellers or day labourers – had gone to work, a team of 50 'goons' with iron bars and knives arrived and poured petrol on houses and set them on fire, and then held off the fire service until police arrived to support the fire fighters. An estimated 600 shanties and 125 shops were destroyed. The daily *New Age* said residents demonstrated against the local MP, Aslamul Haque, whom they blamed for the evictions. 'Aslam's men torched our houses pouring petrol to evict us by force, because High Court has ruled against our eviction,' said Nadim Mohammad, a slum leader. Aslam denied the allegation.[1] Aslam is what is locally called an 'influential person'. As well as being a member of parliament, he is founder and chair of Maisha Property, which bills itself as a 'pre-eminent developer of land in the Dhaka metropolitan area'. Its website says 'When we purchase land, it's always based on strategic location so that we can extend urbanization and create potential sector for real estate and industrial development.'[2] Kallyanpur has become a 'strategic location' because one route of the proposed bus rapid transit system will have a station there, raising the value of the land; Maisha Group wants to build the line and has made a presentation to the prime minister.[3]

The Centre for Urban Studies notes that Dhaka has one of the highest prices of residential land in the world – more than \$12,000 per m² in Gulshan, the most expensive part of the city. In Dhaka two-thirds of the cost of a new apartment is the cost of the land.[4] Thus land grabbing can be hugely profitable.

Probably half of the Bangladesh urban population lives in slums. A survey in 2005 showed that 37 per cent of the Dhaka population was living in 5,000 slums (see Table 10.1). Most slum dwellers live four or five people per room. The Bangladesh

---

1   *New Age* and *Dhaka Tribune*, various articles, 21–23 January 2016.

2   http://maishagroup.com/md-aslamul-haque-mp/ and http://maishagroup.com/business-units/development/maisha-property-development-ltd/, accessed 6 February 2016.

3   http://maishagroup.com/business-units/transportation/brt/, accessed 8 February 2016, and Shahin Akhter, 'Gabtoli-Azimpur BRT, Expressway still in the Woodwork', *New Age*, http://newagebd.net/106104/gabtoli-azimpur-brt-expressway-still-in-the-woodwork.

4   Nazrul Islam and Salma A. Shafi, 'Towards Affordable Housing for Low Income Groups in Urban areas of Bangladesh', *Shelter* 14 (2013): 83.

Table 10.1 Slum populations of the four largest cities, 2005

City	Dhaka	Chittagong	Khulna	Rajshahi
Number of slums	4,966	1,814	520	641
2005 slum population	3,420,521	1,464,028	966,837	489,514
Slums as % of total city population	37%	35%	20%	32%
Slum density, persons per ha	2,202	2,551	1,330	672
% of population in slums				
Up to 100 people	39%	24%	46%	48%
101 to 1000 people	48%	57%	48%	48%
1001 or more people	13%	19%	6%	4%
Persons per room				
Up to 3 people	2%	14%	4%	19%
4 people	32%	25%	50%	43%
5 people	55%	45%	42%	30%
6 or more people	8%	13%	4%	5%

Source: Results from a census and mapping of slums in the six main cities of Bangladesh in 2005.[a]
[a] Nazrul Islam et al., *Slums of Urban Bangladesh: Mapping and Census 2005* (Dhaka: Centre for Urban Studies, 2006) and Gustavo Angeles et al., 'The 2005 Census and Mapping of Slums in Bangladesh: Design, Select Results and Application', *International Journal of Health Geographics* 8 (2009). doi:10.1186/1476-072X-8-32.

Bureau of Statistics (BBS) *Census of Slum Areas and Floating Population 2014*[5] identified 13,938 urban slums, compared to only 2,991 slums recorded in a similar survey in 1997. This is the result of migration, but BBS also argues that 'this change is primarily because the residents of the big slums in Dhaka, Chittagong, Khulna and Rajshahi cities have been evicted. The displaced inhabitants have formed small groups to create small slums.'

Most of the half million people who pour into Dhaka each year usually live first in slums. Each slum is different, but all start on land that was not seen as suitable for proper (*pucka*) buildings. Self-built houses are sometimes constructed beside railways or other unused spaces. In Dhaka this is often over water or in areas that flood. Shanties are built on poles over lakes or floodplains, or on reclaimed land behind new dykes. As noted in Chapter 9, the land is usually controlled by an influential person and below them is a network of *maastans* (musclemen or enforcers) and other intermediaries. The influential person wants to increase the value of the land, so the next step is to bring it above water level by filling with rubbish, sand illegally dredged from rivers and soil from building sites. In Dhaka this can happen literally overnight, with residents evicted and a section of land filled.

5 Bangladesh Bureau of Statistics (BBS), *Preliminary Report on Census of Slum Areas and Floating Population 2014* (Dhaka: Bangladesh Bureau of Statistics, Ministry of Planning, 2015), 21–22.

On the higher land, shanties are replaced with slightly more permanent structures, usually built of bamboo, and often two or three stories. Again, the land is often controlled by an 'influential person' who tries to raise funds for a major development or to sell the land at a high profit. This can take five years or more, so they lease the land to others to construct precarious buildings which will be rented by the room, usually one family per room. Those who build their own shanties often rent out rooms. As the land becomes more valuable, residents are evicted and the area redeveloped. This happens several times – first there are precarious shanties, then when the land level is higher somewhat better one-storey buildings, and then when the fill is more stable, bamboo three-storey buildings. Finally, when the area is established, concrete apartment blocks can be built to be rented or sold to the better off. At each stage residents are evicted, often by force; where they resist, fires are not uncommon.

Some slums, particularly on government land, have an established population who feel they own their self-built homes. They have water, latrines and electricity; there are schools, shops and community organizations. But most slums are precarious. Basti[6] means slum in Bangla and Pora Basti, cited at the beginning of this chapter, means 'burned slum'. It has that name because it was rebuilt after a 2003 fire when an influential person tried to evict the residents. Indeed, what happened in Pora Basti has been repeated countless times in Dhaka, and it underlines the insecure position of most Dhaka residents who know they will be evicted when they cannot pay their rent (e.g. if illness or injury or flood means a man cannot pull his rickshaw) or when the landlord wants to develop the land.

Pora Basti is one of nine slums in Kallyanpur which together cover 20 hectares and house up to 40,000 people – a density of 200,000 people per km^2, which is nearly five times the already high density of Dhaka (see Table 9.1). Another slum in Kallyanpur is Beltola, established more than 30 years ago, and now with some tenure security. Most buildings are bamboo and corrugated iron, but some people have built brick buildings because of the fear of fire. Most are one storey; a few are two storey. Each room houses a family; cooking is done on the narrow dirt streets. Beltola is built on land at the edge of a lake – in this city of lakes, rivers and canals most slum land is carved from flood plains or river banks. There are no high hills here, but there is a small slope and houses go right down to the edge of the lake. When we visit in the dry season, the pond is covered by water hyacinth and looks quite pleasant. But in the monsoon, the water level rises by one metre, flooding all the lower houses. Most residents build a small platform inside the house and live on the platform until the water goes down. The local toilets are on raised plinths to stay above water. Raising a bamboo-framed house requires more bamboo poles to lift it up. 'We cannot afford to raise our houses. Each bamboo pole costs 300 taka [$4] and we do not have enough money' explains Monoara Begum, a community leader sitting in her tiny shop in Beltola. In fact, Kallyanpur residents are not so poor. Our survey there showed an average monthly income of 10,463 taka ($150) per month.

---

6 Many Bangla words do not have agreed English spellings; *basti* is also written *bosti*, *bustee* and *bastee*.

Informal or Slum?

The term 'slums' is controversial as it is often seen as pejorative. Alternative terms are informal or low-income settlements. But 'slum' is used by the United Nations, both by Habitat and in the Millennium Development Goals (MDGs); the major umbrella organization is Slum Dwellers International; the word is used by the Bangladesh Bureau of Statistics (BBS) and is commonly used by Bangladeshis. So we use slum here. Habitat says that slums are characterized by five attributes: 1) overcrowding and high density 2) insecure tenure and irregular or informal settlements 3) unhealthy living conditions and hazardous locations 4) lack of basic services and 5) poverty and social exclusion.[7]

The BBS[8] defines four kinds of house and the 2011 census[9] gave this distribution in the Dhaka urban area:

- *Pucka*: floor, wall and roof are made of cement, bricks and stones – 39 per cent.
- *Semi-pucka*: walls are generally made of cement and bricks and roof is made of tin, asbestos sheet, wood or bamboo – 34 per cent.
- *Kutcha*: walls are made of clay, wood, bamboo, straw or raw bricks and roofs are made of tin, bamboo or straw – 25 per cent.
- *Jhupri*: much smaller and lower than a general house (to enter into the house one has to bow down his/her head) and made of straw/leaf, polythene, bamboo or tin – 2 per cent.

For all urban areas, about one-third each of homes are pucka, semi-pucka and kutcha. Only one-third of families live in proper houses, which suggests that somewhere between one-third and two-thirds of urban Bangladeshis live in slums. Habitat (Table 9.1) puts the national figure at 55 per cent of the urban population living in slums.[10] The World Bank says that around 62 per cent of urban residents are living in informal settlements or slums.[11] Whatever the actual percentage, it's a lot of people in very poor living conditions.

---

7 Michael Kinyanjui, 'Development Context and the Millennium Agenda: The Challenge of Slums', revised and updated 2010 from *Global Report on Human Settlements 2003* (Nairobi: United Nations Human Settlements Programme – UN-Habitat, 2003), revised and updated version: http://unhabitat.org/wp-content/uploads/2003/07/GRHS_2003_Chapter_01_Revised_2010.pdf.

8 BBS, *Report on Rural Credit Survey 2014* (Dhaka: BBS, Ministry of Planning, 2014), 19–21.

9 BBS, *Bangladesh Population and Housing Census 2011, National Report, Volume – 1 Analytical report* (Dhaka: BBS, Ministry of Planning, 2015), table H02, 184–87.

10 Habitat, Statistical Annex to World Cities Report 2016, table 2.3. Accessed 10 March 2016, http://unhabitat.org/urban-knowledge/global-urban-observatory-guo/.

11 World Bank, 'Bangladesh – Pro Poor Slums Integration Project: Environmental Management Framework' (Dhaka: National Housing Authority, 2014), i. Accessed 12 September 2016, http://documents.worldbank.org/curated/en/2014/07/20163472/.

Table 10.2 Percentage of each type of household to have access to services

	Water			Electricity
	Tap	Tube–well	Other	
Pucka	89%	10%	0%	100%
Semi-Pucka	53%	46%	1%	98%
Kutcha	17%	78%	5%	79%
Jhupri	36%	49%	14%	70%
Sanitation	Flush toilet[a]		Sanitary non-flush	Non-sanitary or no toilet
Pucka	76%		23%	1%
Semi-Pucka	37%		57%	6%
Kutcha	15%		48%	37%
Jhupri	9%		36%	55%

[a] A flush toilet is defined in the census as a water-sealed toilet. Most urban non-sanitary toilets are pit latrines or hanging toilets – which literally hang out over a pond or canal.
Source: BBS, *Bangladesh Population and Housing Census 2011*, tables H02 and H03, 184–89.

In Dhaka, only 36 per cent of people own their house[12] while 61 per cent rent.[13] The BBS conducted a slum survey in 2014 and found that 84 per cent of slum households have a mobile phone, 79 per cent have a fan, 48 per cent have television and only 7 per cent have a refrigerator.[11] Among productive assets, 6 per cent have a rickshaw, 4 per cent have a sewing machine and 5 per cent have a bicycle.

There is a significant class stratification in the slums. A survey by Nicola Banks in four slums in 2009 found three levels.[15] At the top is a slum elite with political or business links outside the slum, and who, through their connections, are able to access resources and opportunities. They can secure good jobs and move out of poverty, even though they are still living in the slum. This group also rents out rooms (which can be quite lucrative) and controls local services such as selling electricity or water. A middle-tier has family, kinship and political links to local leaders. These two groups are house owners. But the majority of low-income households are at the bottom of these hierarchies, and through exploitative patron-client relationships are dependent on the slum elite for their access to shelter,

---

12 'Owning' a house covers a range of situations. A few people actually have some sort of document. More commonly, they are recognized as the original squatter who occupied the land and built the house – or that they bought the house from the original occupant. Recognized ownership means they cannot be evicted by local thugs or intermediaries, and will not be evicted from state land. But they could still be evicted from private land.
13 BBS, *Bangladesh Population and Housing Census 2011*, table H01, 183.
14 BBS, *Preliminary Report on Census of Slum Areas.*
15 Nicola Banks, 'Urban Livelihoods in an Era of Climate Change: Household Adaptations and their Limitations in Dhaka, Bangladesh', in *Urban Poverty and Climate Change*, ed Manoj Roy et al. (Abingdon, Oxon: Routledge, 2016): 113–29.

services and security. In her survey, Banks found that less than a third of households own their own house, but of the home-owning group, nearly 70 per cent have other productive assets such as rental rooms, shops, rickshaws or sewing machines. In contrast, only 12 per cent of tenants have productive assets.

The BBS survey showed the highest percentage of slum dwellers (17 per cent) reported rickshaw pulling as their main source of income, followed by general business (16 per cent), garment worker (14 per cent), service (14 per cent), construction worker (8 per cent), day labour or porter (8 per cent) and transport worker (8 per cent). Many do some of their work at home or on the street next to their home – recycling plastic or paper, tailoring, making goods they can sell or doing repairs, providing services, or at least storing goods from street trading. Thus some shacks and semi-pucca houses have one room to live in and a second as a workroom. Home-based work is particularly important for women, who often sell their product within their own neighbourhood.[16]

The extreme poor are usually defined as those earning under 5,000 taka per month ($60 per month, $2 per day). One third of their income goes on rent; the rest is spent on food, and they often cannot afford enough to eat. Data is poor so it is difficult to estimate what share of slum dwellers are extremely poor. But one survey showed that the poorest 20 per cent had an income below $30 per month ($1 per day) and another survey showed that nearly three quarters of residents were in debt, often to money lenders.[17]

Rents in the slums start at 1,000 taka ($13) per month for a room and go up to 3,500 taka ($45) or more. This means Dhaka's slum dwellers pay more that the city's middle class, notes Khurshid Zabin Hossain Taufique, deputy director of the government's Urban Housing Directorate. Slum dwellers pay rents starting at about $1.50 per month per m^2, whereas serviced flats in middle-class blocks start at $1 per month per m^2.[18]

NGOs Fill the Gap

Table 10.2 shows that poor people have less access to safe water – usually a common tap serving several houses or a locally drilled tube well – and are much less likely to have proper toilets, although nearly everyone in Dhaka has enough electricity to power a light bulb and charge a mobile phone.

Dhaka Water Supply and Sewerage Authority (WASA) is responsible for water supply, sewerage disposal and storm water drainage in the megacity. But its drains cover only 38 per cent of the city, and only 30 per cent of households are linked to a sewerage system; most household have septic tanks. WASA Managing Director Taqsem A. Kahn admits that 'a significant number of the more vulnerable LIC [low income community] dwellers often have minimal or no access to water services'. In part this is because WASA will not deal with communities it says are on 'illegal lands'. Many poor people have to buy

16   Shahadat Hossain, *Urban Poverty in Bangladesh* (London: I. B. Tauris, London, 2011), 153–54.
17   Nicola Banks, *Urban Poverty in Bangladesh: Causes, Consequences, and Coping Strategies*', (Manchester, UK: Brooks World Poverty Institute, 2012), BWPI Working Paper 178.
18   Jemima Rohekar, 'Slums on Rent', *Down to Earth* (New Delhi, India), 16–31 March 2016. Accessed 13 September 2016, http://www.downtoearth.org.in/news/slums-on-rent-53291.

water from vendors. So to expand water supply to the slums, WASA is working through local and international nongovernment organizations (NGOs) who guarantee payment of the bills.[19]

Part of the Bangladeshi version of neoliberalism is that social services are partly carried out by NGOs rather than the state. Two of them, Grameen and BRAC, have become major multinational corporations and aid agencies in their own right. Both were set up after independence and stressed microcredit; profits from that lending in the 1990s were important in financing their expansion. Muhammad Yunus founded Grameen Bank and he and the Bank were awarded the Nobel Peace Prize in 2006. By the end of 2015 Grameen Bank had $2.4 billion in deposits. Grameenphone is the largest mobile phone operator in the country. Grameen Trust now has microcredit and social business projects in 41 countries, including the UK. BRAC, originally the Bangladesh Rural Advancement Committee, says it is the largest nongovernmental development organization in the world, in terms of number of employees. In 2014 it spent $932 mn, including nearly $200 mn from international donors (46 per cent from DfID – UK Aid). BRAC had projects in 11 countries. It runs a university and a bank with deposits of $1.6 bn.

The role of NGOs as service providers and intermediaries has become problematic. As we noted in the previous chapter, local NGOs have become part of Bangladesh's patron-client political system. This means they do not promote autonomous, rights-based, campaigning community organizations, but rather community-based organizations (CBOs) that fit within the political system. Water is a good example, where NGOs such as DSK create their own CBO and organize the water contract between WASA and the CBO. But in the patron-client atmosphere, this CBO is not controlled by the community and remains subordinate to the parent NGO.

In other countries, including neighbouring India, NGOs have promoted the development of more activist CBOs. In India, the National Slum Dwellers Federation (NSDF) and the Society for the Promotion of Area Resource Centers (SPARC) are an alliance of CBOs in more than 10,000 slums in 70 cities, with the specific goal of 'mobilization and organization of communities of the urban poor'. In contrast, attempts to form federations of slum CBOs in Bangladesh have been resisted and have failed.

There are repeated complaints that NGOs compete with each other and that they refuse to work through community organizations. Many Bangladeshi NGOs are built on microfinance, and they are often accused of working first with their borrowers. There is a particular problem in post-flood relief and rehabilitation where each NGO has its own programme and does not listen to local people and often does not help the most needy.[20] In part this reflects the scramble for funds from international donors or agencies. Local NGOs have to move quickly to write proposals and obtain grants, which leaves little time for participation from local stakeholders. Indeed, while international donors pay lip

19 Taqsem A. Khan, 'Dhaka Water Supply and Sewerage Authority: Performance and Challenges', Dhaka Water Supply and Sewerage Authority, 2012. Accessed 12 September 2016, http://dwasa.org.bd/wp-content/uploads/2015/10/Dhaka-WASA-Article-for-BOOK.pdf.

20 Ainun Nishat et al., *The 1998 Flood: Impact on Environment of Dhaka City* (Dhaka: Department of Environment, Ministry of Environment and Forest, 2000), 227.

service to community 'voice', in practice they dismiss grassroots knowledge in favour of that of professional experts who often know little of the Dhaka slums.[21] The scramble for funds also means following international fashion and what looks good in photographs, so NGOs will continue to install water pipes but not maintain them. There is no thought of depreciation and the life of a system, which can be less than ten years for communal toilets. But after three or five years, when the money was finished, the original NGO had already moved on. So the community waits until another NGO gets money to do the same thing again.

Government health initiatives tend to be more rural than urban,[22] and primary health care in urban areas is contracted out to NGOs.[23] In the slums most people use private and informal medicine sellers and other health care providers, in part because NGO services tend to be open only when people are working. NGOs facilities are largely outside the slums and clustered in a few places rather than spread out across the low-income areas.[24]

In a thoughtful study, Enamul Habib, a senior assistant secretary in the Bangladesh Civil Service, looked at NGOs in slums in Dhaka.[25] He noted that 'the major NGOs are more concerned with rural development' and when they do work in Dhaka, 'NGOs are not keen to get involved in housing,' which is the main need. He concluded: 'Many NGOs choose their field of intervention according to donor guidelines, which are unlikely to reflect the actual slum needs. NGOs' work is poorly co-ordinated, with relatively little emphasis on organising communities to manage their own problems. [...] NGOs should then focus on the problems which are more important to the slum dwellers, rather than on the need to attract foreign donations.' And a study in Khulna showed that 'money and technical assistance always comes from the top (nationally or internationally) and are allocated using a top-down method. There is no control or participation by the community representatives in deciding how the money is allocated.'[26]

Nevertheless, some services are better than none, and where NGOs work, the local residents are appreciative. Monoara Begum runs a shop in Beltola, Kallyanpur, adjoining Pora Basti described at the start of the chapter. Her shop is on the main street of the settlement, and she says that 'the street used to flood, but it does not any more because an NGO built a storm drain.' With outside help, improvements can be made (see Box 10.1).

---

21 Nicola Banks, Manoj Roy and David Hulme, 'Neglecting the Urban Poor in Bangladesh: Research, Policy and Action in the Context of Climate Change', *Environment and Urbanization* 23 (2011): 487–502.

22 Farid Khan, Arjun S. Bedi and Robert Sparrow, 'Sickness and Death: Economic Consequences and Coping Strategies of the Urban Poor in Bangladesh', *World Development* 72 (2015): 255–66.

23 Alayne M Adams, Rubana Islam and Tanvir Ahmed, 'Who Serves the Urban Poor? A Geospatial and Descriptive Analysis of Health Services in Slum Settlements in Dhaka, Bangladesh', *Health Policy and Planning* 30 (2015): 132–45.

24 Adams, Islam and Ahmed, 'Who Serves', 142.

25 Enamul Habib, 'The Role of Government and NGOs in Slum Development: The Case of Dhaka City', *Development in Practice* 19 (2009), 259–65.

26 Manoj Roy, Ferdous Jahan and David Hulme, '*Community and Institutional Responses to the Challenges Facing Poor Urban People in Khulna, Bangladesh in an Era of Climate Change*', BWPI Working Paper 163 (Manchester, UK: Brooks World Poverty Institute, University of Manchester, 2012): 51.

Monoara Begum and her daughter Ayesha in her tiny shop in Beltola, Kallyanpur, Dhaka. Photo: Joseph Hanlon.

DSK installed the water and toilets in Pora Basti and Beltola, but left it for the community to maintain, which it could not afford; more recently, another NGO had to come in to rehabilitate the system.

But ten-storey blocks of flats are already encroaching on the Kallyanpur slums. How long before Begum is evicted or her little shop burned down? Unpredictability makes people unwilling to invest time or money. People who are unsure where they will live next year are less likely to take part in local politics or to invest in upgrading their homes. NGOs will not work in Kallyanpur if they think their pipes and toilets will be destroyed and the residents evicted before their project is finished. As part of its Pro Poor Slums Integration Project in five small secondary cities, the World Bank agrees that 'solutions to insecure tenure are critical to the sustainability and equity of slum developments'. It proposes a model where 'both the land owner and the community people share the land and reach to a win-win situation after negotiating on this issue' and it also proposes another model with $75 mn to 'allow for these community groups to receive a mixture of loans and grants to assist with securing tenure (through lease or purchase of land)'.[27] But in Dhaka, where land is exorbitantly expensive and the land under Monoara Begum's tiny shop might be worth $25,000 or more, it would take massive World Bank loans to buy the land, and these seem unlikely to be on offer; land owners are unlikely to be willing to share unless forced to. This presents real obstacles to climate change adaptation and also partly explains why Dhaka is such an unlivable city.

At least the World Bank is considering tenure. In a study in Chittagong, Bangladesh's other major city, Ronju Ahammad pointed to a UK-funded project which only worked

---

27 World Bank, *Bangladesh – Pro Poor Slums Integration Project: Environmental Management Framework* (Dhaka: National Housing Authority, 2014), i.

with people with secure tenure, who set up the community development committee. Thus the project excluded the poorer landless renters. The project stressed improving water and sanitation, 'however at this stage people would be more interested in securing tenure'.[28]

Later in this chapter we look at what people who feel secure are doing to adapt, and what that might suggest for the future. Slums may look like landscapes of disaster, but they are also places of hope and aspiration; people have come to the megacities to better their lives and to build better futures for their children. They do not want to live on land contaminated with untreated sewage deposited during monsoon floods, but they have no choice. We have seen that with support and security, local people can improve their environment. Effective collective action is possible but this will require institutions that actually support Dhaka's poor majority, in part by reducing the risk of eviction. Three things are needed: security of tenure; municipal supply of basic services such as water, sewerage and rubbish collection; and support for autonomous community organization and networks of CBOs rather than NGO dependency.

Box 10.1　Services in Kallyanpur

Dushtha Shasthya Kendra (DSK) is a medium-sized NGO. Its microfinance deposits are only $11 mn and its budget is only $7 mn per year. Almost half of this goes to water and sanitation ($3 mn), with projects in 30 Dhaka slums, including in Kallyanpur, where a water system was installed in 2004 serving 21,000 households. First, DSK set up a water group of local women, then it signed a bulk contract with WASA and guaranteed to pay. Next step was to install water pipes. DSK supplied the pipe and supervised construction but the community installed it. In Kallyanpur a tap serves 30 to 70 families. Toilets and septic tanks were also installed. But over the years, lack of maintenance has meant that many stand pipes (public taps) became dilapidated, and in 2015 other NGOs, Water and Sanitation for the Urban Poor and NGO Forum, entered Kallyanpur to repair a number of existing water points and sanitation chambers. But there are problems with alleged corruption in the management committee and failure of the residents to pay their monthly bills of 98 taka ($1.40).

Kallyanpur's toilets are not connected to any sewage system, so the waste eventually drains down to the two ponds (*jheels*)[29] – particularly from septic tank overflows and during heavy monsoon rains and flooding. The ponds are highly contaminated with faecal matter, but some residents continue to wash their clothes and cooking utensils in the ponds. Testing of plants and food in Kallyanpur indicated high levels of phosphorous contamination, associated with human waste, as

---

28　Ronju Ahammad, 'Constraints of Pro-Poor Climate Change Adaptation in Chittagong City,' *Environment and Urbanization* 23 (2011): 512.

29　The terms *beel* and *jheel* overlap as both are natural basins, but in common use a beel is dry for part of the year and a jheel is a permanent pond or lake.

well as E.coli, coliforms, aerobic bacteria, Staphylcoccus aureus, yeast and moulds. This has major health implications for residents (especially children), who grow and consume plants within the settlement and fish from the beels.

One of the authors tested the water quality (for coliform) and found it was relatively safe to drink at the tap. However, upon testing quality of the same water at the point of consumption in the household (often just 100 m from the water point) significant levels of contamination, including over 1,000 times the levels of faecal coliform, were found. The local environment is so contaminated from overflowing drains, leaking sewage and the monsoon floods that people handling water containers contaminate them, making the water unsafe for human consumption. This can explain the prevalence of some diseases (such as cholera, typhoid and diarrhoea), especially during rainy season. Nevertheless, the limited toilet system has replaced the previous open pits and toilets hanging over the beels; 86 per cent of households use a reasonable toilet which has improved health, although 18 families share each toilet.

In 2005, DSK and the local community groups constructed a drainage system in the slum. But it was not adequately designed – the pipes were too narrow and there was a dip in the middle of the pipe which disrupted the flow, while outlets were connected to the ponds increasing contamination of water bodies. No maintenance was organized and waste and rubbish clogged the drains, exacerbating localized flooding. The local ponds, in turn, drained out towards a canal, but a private hospital was built which cut off the channel, leading to flooding. The hospital refused to negotiate. So in 2010 on a long holiday weekend, a hospital worker who lived in the slum unlocked the gates and allowed the community to install a pipe across the hospital site to restore the drainage. The community action was effective, but it also underlined the lack of public action by the city.

Nevertheless there are still problems. A World Bank study of Kallyanpur said that just to deal with current flooding would require pumps and new, larger drain pipes costing $5 mn.[30] Dealing with the additional rainfall expected from climate change would require additional pumps.

Since 2001 there has been an electricity system serving 1,000 shacks in Kallyanpur, about one quarter of the households. It takes power illegally by tapping into the electricity lines. A 15-member committee collects monthly bills and maintains these connections. The cost is 250 Tk ($3.50)/month for a light, fan and television, and 300 Tk/month for a fridge.

No Return to the Countryside

Shahadat Hossain points to the high standards of living of the better off in Dhaka, and argues that 'it may well be the only megacity in the world where the inequality between

---

30 Susmita Dasgupta et al., *Urban Flooding of Greater Dhaka in a Changing Climate Building Local Resilience to Disaster Risk* (Washington: World Bank, 2015), 130–32.

the rich and the poor is so high'.[31] The Dhaka elite still believe that the poor majority have no permanent place there and can be moved at will to allow housing and profits for the better off. But these millions of slum dwellers are not going to move 'back' to the countryside. They are now integrated into the economic life of the city, which depends on their labour. They have children and increasingly grandchildren who have no rural place to go 'back' to. Enamul Habib makes the obvious point that 'the slums need to be considered as part of the city'.[32]

There is a stark difference in attitudes between rural areas (see Chapters 4–6), where there has been major movement on responding to environment and climate change, and urban areas, where exploitation of land and water is given top priority with no concern about planning physical development or climate change. Drains, rivers, lakes and flood plains are filled. A BRAC University survey in 2012 showed that in Dhaka slums, 35 per cent of families threw household waste directly into drains and 32 per cent dumped it along roads or on vacant land.[33] A diplomat told us that embassy staff saw a municipal rubbish lorry dumping the waste into a pumping station canal – and buildings were already being built on the filled land. Land is grabbed for apartment blocks and factories; the remaining rivers are polluted with industrial effluent and sewerage. Huge profits are to be made, and the price is paid in ever-worse flooding. The elite in their tower blocks are above flood level, but when the Economist Intelligence Unit finds Dhaka the least liveable city in the world, except for ones destroyed by war, perhaps a few more people might consider a change in strategy.

The poor will be worst hit by climate change. Their homes are closest to water level, and without drains and raised construction, they will flood first. Flimsy shanties will be destroyed even in minor cyclones. Rickshaw pullers must work in heat and rain, and cannot stay in air-conditioned offices. They cannot afford to 'adapt' to climate change; the best they can hope is to cope with the damage it causes. There are many ideas about how to adapt to climate change, but the poor cannot afford them. The slum-dwelling poor are at least half of Dhaka; discussion of urban impacts and adaptation starts with them. Of course, the biggest problem is that the poor have little money and cannot afford to adapt to exiting environmental pressures, which will be made worse by climate change. We are not looking specifically at urban poverty here; it has been dealt with by the authors elsewhere, in the 2016 book *Urban Poverty and Climate Change*.[34]

---

31 Hossain, *Urban Poverty*, 16.
32 Habib, 'The Role of Government and NGOs', 263.
33 Liaquat Ali Choudhury, 'Quality of Urban Life: Service Realities', in *Bangladesh Urban Dynamics*, ed Hossain Zillur Rahman (Dhaka: Power and Participation Research Centre, 2012), 147–69.
34 Manoj Roy et al., *Urban Poverty and Climate Change* (Abingdon, Oxon: Routledge, 2016). See also various working papers of the Brooks World Poverty Institute of the University of Manchester: http://www.gdi.manchester.ac.uk/research/publications/working-papers/archive/ bwpi.

The Changing Slums

In the first years after independence, squatters in Dhaka and Chittagong occupied government-controlled land, for example, along railways or land deemed not appropriate for development, and gained some permanence. Dhaka's biggest slum is probably Korail, home to up to 200,000 people. Disputed between two government departments and not built on, the land was squatted in the 1980s.[35] Many of the residents work as cleaners or servants in the nearby high-end commercial and residential areas of Gulshan and Banani.

In recent times, low-income settlements have been on private land[36] – often with a title acquired by an influential person by dubious means. Often it is low-lying land on a flood plain or behind a new dyke. The land is claimed by allowing people to build shanties and temporary structures. Normally the landowner appoints middlemen – often mastaans or musclemen – or rents out part of the site to third parties who build basic dwellings to rent out. Rent is collected weekly from individual slum dwellers. The landowner usually controls the supply of water and electricity (often from a generator).

Surveys show that threat of eviction is the biggest worry for most slum dwellers.[37] Income and poverty-related challenges come next. Our survey in Kallyanpur showed the top problems expressed by men and women were fear of eviction, crowded living conditions and insufficient toilets. For women, lack of water and poor drainage came next, while for men it was lack of jobs.

A study in Korail found that residents do not recognize climate change, but they do recognize climate variation.[38] They report increased heat, and with it increasing disease. They also recognize flooding as a major issue. There is a serious attempt to adapt to these climatic factors, which are exactly the issues that will be made worse by climate change. To reduce heat, residents adopted techniques from their rural origins. This includes growing vines to cover the roofs, building around shaded courtyards and creating either a false ceiling or canopy over the roof, which allows air flow through the room. In houses with corrugated iron roofs, people put in some insulation made from newspaper or plastic. During hot periods, most households use a fan. Electricity is not metered and instead people pay vendors according to the number of electricity points. Most have just one, and in the hot season use a fan instead of a light.

In Korail, over time, people try to move up the slight incline away from the lake to try to reduce the likelihood of flooding. Those near the lake try to raise their shanties on stilts (which also allow air flow underneath, also reducing the temperature). Furniture is also put on two or three bricks to try to raise it above normal flood level. During flood, platforms are often constructed inside the house, so people can sleep and cook above

---

35 Huraera Jabeen, Cassidy Johnson and Adriana Allen, 'Built-In Resilience: Learning from Grassroots Coping Strategies for Climate Variability'. *Environment and Urbanization* 22 (2010): 419.

36 Manoj Roy and David Hulme, 'How the Private Sector Meets the Demand for Low-Income Shelter in Bangladesh', *Shelter* 14 (2013): 90–98.

37 For example, Jabeen, Johnson and Allen, 'Built-In Resilience', 420, and Banks, Roy and Hulme, 'Neglecting the Urban Poor in Bangladesh,' 496.

38 Jabeen, 'Built-In Resilience', 422–26.

Notunbazar, Dhaka, has moved into a second phase with multistorey but still precarious structures. On this street, the ground floors of buildings are made of brick to which an upper floor of bamboo and metal sheets has been added. With no internal staircase, the only access is by an external ladder. Photo: Joseph Hanlon.

water level. Further up the slope the issue is flooding from heavy rain. Communities dig drains, and individuals put a 10 cm high board across the doorway to serve as a barrier to rainwater. If nothing is done to upgrade these slums, then climate change will force residents to do more of this kind of adaptation.

Building Better Housing for the Poor

However, the process of raising and developing land has created some space for innovation by private developers. When land has been filled, it is often rented out for periods of up to five years. New temporary developers are often themselves of modest incomes, but to pay the rent and make a profit, they must build two- or three-storey buildings with large numbers of rooms to rent. These are still temporary, but they are semi pucka with a

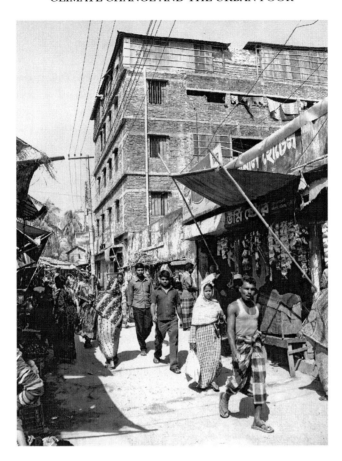

This well-built five-storey block on the main street of Moti Jharna, Chittagong, is entirely unauthorized but provides decent accommodation, with water, toilets and cooking areas. Photo: Joseph Hanlon.

planned five-year life. Many are dank, dreary and unpleasant; the rooms are small, with space for little more than a double bed, and rents start at $5 per week. But some have rooms a bit larger and introduce innovative design practices, such as cement on bamboo frame flooring and communal toilets and ventilated cooking areas within the building.

Some landowners now build pucka structures of up to six stories. These are still seen as temporary, because the land has more value than the building, but they have a possible life of 25 years or more. They are totally unauthorized and do not meet any building codes, but can be of quite high standard.[39] The best are similar to the low-income housing constructed by the Peabody Trust in London in the 1870s, just as London was going through its expansion to become a megacity. These houses have electricity, water, toilets, sewerage, cooking areas and some small communal space. A typical room is 6 to 12 m^2

---

39 Roy, 'How the Private Sector'.

Box 10.2 Less dangerous than a Chittagong bus

Moti Jharna slum winds up a long narrow valley between steep sand hills. The main street is a row of market stalls, selling a wide variety of vegetables and grains. There are goats and cattle wandering the street and even a few cars. This is not a poor area. Unlike Dhaka, which is flat, Chittagong lies in the bottom of Bangladesh's only hills, but there are regular landslides, including one triggered by 425 mm of rain on 11 June 2007, which affected five slums and killed 127 people.[40] Chittagong City Corporation considers Moti Jharna a landslide risk and does not officially permit construction there. So it is entirely informal and unauthorized – although some people have obtained fake land registration titles from corrupt official and politicians. This slum is in transition. Along the bottom of the valley are pucca and semi-pucca buildings, up to five storeys high. This street has drains, water and electricity.

Going up the steep hillsides, the buildings are more temporary. Steep steps up the hill are sometimes made of sandbags. Water has to be taken up the hill in buckets from taps on the main street. Some cement drains have been constructed but others are just cut into the sand because downhill residents have not allowed the construction of concrete drains. Where one had been built, we saw a woman dumping her rubbish into the cement drain running in front of her house. This blocks the flow and sewerage flows around the houses. It is hardly surprising that residents complain about drains, sanitation and rubbish.

At the top of one staircase, Kalim and his wife have built a semi-pucca house. They rent out rooms, probably their main source of income. They invite us into the large single room they live in. It is comfortable and feels like a permanent home. As well as the double bed (with mosquito net), there is electricity, a refrigerator and a fan. The kitchen is behind, with a water filter. Kalim has built a latrine on the hillside on the other side of the path.

But there is erosion. The corner of his house is already propped up by sandbags, filling the space where the hillside had been eroded. Moti Jharna is considered a 'high-risk area'.[11] We ask Kalim if he is afraid of a landslide. He replies: 'Taking buses in Chittagong is more of a risk.'

Being poor in Chittagong means taking risks. In 2012 Chittagong received 400 mm of rain in one day and 90 people were killed in the landslides. Climate change will bring more intense rainfall and stronger cyclones. How long will Kalim's house, and the others on this hillside, survive?

---

40 Ahammad, 'Constraints of Pro-Poor,' 507.

41 Amanullah Bin Mahmood and Mamunul H. Khan, 'Landslide Vulnerability of Bangladesh Hills and Sustainable Management Options: A Case Study of 2007 Landslide in Chittagong City', Proceedings: SAARC Workshop on Landslide Risk Management in South Asia, 11–12 May 2010, Thimpu, Bhutan, 67.

Kalim and his children and their friends stand on the steep stairway beside his house on the hillside of Moti Jharna, Chittagong. Erosion is a serious problem and sand bags already prop up the corner of the house and form part of the steps. Photo: Joseph Hanlon

(of which a double bed takes up 2 m^2) and houses one family or up to six single men. Sometimes account is taken for many slum dwellers being self-employed and some rooms are rented for commercial or productive use, such as tailoring. Rents are higher and tenants are lower middle class by slum standards. Usually there are communal toilets and cooking areas, but the next step up the social ladder is to have a private toilet and cooking area in the room, separated off by a curtain. Sometimes there is even proper flood and rainwater drainage, which is a first step towards adapting to climate change.

Dhaka and Chittagong have building regulations, but they are not enforced, and seem not to apply in areas where building permission has not been granted. A bribe or being an influential person is all that is needed. As Rana Plaza (Box 9.1) showed, modern buildings that look sturdy and are outside the slums can still crumble. There are no controls on unauthorized building. The possibility of collapse is real. The earthquake in Manipur, India, near the borders with Bangladesh and Myanmar on 4 January 2016 damaged many new buildings, raising questions about building codes that do not take account of earthquakes but are, in any case, often ignored. Little noted is that Bangladesh is in an earthquake zone and there have been serious earthquakes. The 1762 quake near the coast of Chittagong caused severe damage in both Chittagong and Dhaka and triggered a tsunami: 'At Dacca, in the kingdom of Bengal, the consequences have been terrible: the rise of the waters in the river was so very sudden and violent, that some hundreds of large country boats were driven ashore, or lost, and great numbers of lives lost in them,' according to a contemporary report.[12] Three tectonic plates meet in Bangladesh.[43] The Indian Plate moves gradually northeast pushing under the Eurasian Plate, causing the Himalayas to rise, and this caused the April 2015 earthquake in Nepal, which killed over 8,000 people. To the east, the Burmese Plate pushes west over the Indian Plate, creating the highlands east of Chittagong. The 2016 earthquake in Manipur, India, was near the meeting of the three plates. Both the Nepal and Manipur earthquakes were felt in Dhaka; although they did not do significant damage, they did provoke discussion about the lack of earthquake preparedness. The director of the Dhaka University Institute of Disaster Management and Vulnerability Studies, Khondoker Mokaddem Hossain, estimated that an earthquake as strong as the 2015 one in Nepal would cause 40 per cent of Dhaka buildings to collapse – in part because the building code is not enforced.[44] Earthquakes and tsunamis are not climate change issues, but any climate change adaptation in Dhaka should also take into account the high earthquake and tsunami risk.

Conclusion: Dhaka Is Third Most Vulnerable City

Dhaka will be hit by climate change. Increased heat will make it harder to work and travel. The combination of rising sea level and increased rainfall means more and more serious floods. Cyclones will be more damaging. The Intergovernmental Panel on Climate Change (IPCC) says that Dhaka is the third most vulnerable city in the world to coastal flooding (after Kolkata and Mumbai) due to sea level rise and worse storm surges.[45]

Banani Lake in Dhaka is surrounded by the diplomatic quarters and some of the most expensive housing in the city. The high-rise blocks of BRAC University overlook the lake,

---

42　William Hirst, 'An Account of an Earthquake in the East Indies, the Two Eclipses of the Sun and the Moon, Observed at Calcutta: In a Letter to the Reverend Thomas Birch, D. D. Secret. R. S. from the Reverent William Hirst, M. A. F. R. S.', *Philosophical Transactions* (London: The Royal Society) 53 (1763): 256–62.

43　'Earthquake Risk in Bangladesh', American Museum of Natural History Collect Curriculum Collections, New York, 2013. Accessed 13 February 2016, http://www.amnh.org/explore/curriculum-collections/earthquake-risk-in-bangladesh.

44　Tapos Kanti Das, 'No Quake Preparedness: People at High Risk', *New Age*, 6 January 2016.

45　Intergovernmental Panel on Climate Change (IPCC), *The IPCC's Fifth Assessment Report: What's in it for South Asia* (London: Climate and Development Knowledge Network, 2014) 11, citing Susan Hanson et al., 'A Global Ranking of Port Cities with High Exposure to Climate Extremes', *Climatic Change* 104 (2011): 100. doi 10.1007/s10584-010-9977-4.

Luxury flats in Banani, Dhaka, look out over the lake – and at the shacks built along the lake shore. Photo: Joseph Hanlon.

but they also overlook Korail slum, which occupies part of the lake shore. A footpath runs around the north end of the lake in Banani; on one side of the path are luxury high-rise buildings, while shacks are built on the other side, on poles over the lake. Goats and chicken wander around, families are sorting paper and plastic, there are rickshaws in the yard, and it is all less than one metre above water level; in heavy rains, this area floods, but it will be hardly noticed in the Banani apartment blocks. The ground floors of these blocks are only used for parking and storage. In the event of a few days of flooding, the better off can remain in their upper storey flats and send their servants out to wade through the water to buy food.

But the majority of slum dwellers already live on the margin. They have adapted as best they can to heat and heavy rains, but in a serious flood they will lose homes and livelihoods. Climate change will make their position much worse but there is nothing more they can do privately to adapt. Many in the upper classes will be happy to see the nearby slums washed away. But the poor majority in Dhaka cannot and will not be carried away in the flood waters.

The distinction between climate and climate change is important. When a family builds its shanty on poles to stay above water level, it is not adapting to climate change but to ordinary flooding. Climate change has not yet had a noticeable impact, but it definitely will – making all of the present climate problems worse. Deeper and more frequent floods, stronger cyclones, and heavier rainstorms will happen. The poor are already adapting to all of the environmental stresses; their carbon footprint is miniscule and they have done the least to cause climate change, but they are the least able to adapt because of all the compounding factors of poverty. This means there is little point in donor and NGO projects targeting climate change and the poor, even though that is the current fashion. The first step is to target poverty and especially the vastly unequal power relations between rich and poor in Bangladesh's cities. Public action is needed to tackle these problems.

The key rural-urban differences are about political power, which is why the urban problem is so intractable. Slum dwellers will need at least two kinds of power – more security of tenure and more political access, for example, through the kind of decentralization such as in rural areas where a councillor represents 4,000 people instead of 120,000, and thus is known and, at least partially, accountable.

Some shift in power would allow Dhaka to look more closely at the innovation and adaptation strategies of the urban poor, and of NGOs and developers working with them, to upgrade and develop the slums for the people who live there, and in ways that will allow Dhaka to respond better to climate change. At the very least, supporting collective action of slum dwellers to improve their slums requires two things: risk reduction, especially reducing the fear of eviction, and providing basic services of water and sanitation that will allow slum dwellers to live healthier and productive lives. This, in turn, will create the space for local people to improve their living conditions and begin to make essential adaptations to climate and climate change.

None of the adaptations discussed in this book, including cyclone shelters and tidal river management, are adaptations to climate change. Instead, they are adaptations to the existing climate. But they are important because they provide the experience and knowledge base to adapt to climate change. Having a living delta and capricious climate has given Bangladesh a head start in understanding how to deal with climate change – both mitigation and adaptation. That is why Bangladesh is keeping its head above water and is taking a global lead on this issue at international meetings.

In rural areas, a more representative form of patron-client politics has empowered the majority and created a working relationship that has led to tidal river management, cyclone shelters and improved rice varieties. There the adaptation that is needed for climate change is already underway.

But in the privatized cities of Dhaka and Chittagong, the poor majority is virtually powerless – both politically and in its inability to respond to climate change. Slum residents have shown huge ingenuity in adapting; there have been some successful experiments in private low-income housing. But the elite have so far refused to cede any power or create any political space to integrate the poor majority into the life and policy making of these two key cities. So far, the influential people sitting in their upper floor offices and flats feel they can ignore what happens on the ground; traffic jams now and climate change in coming decades are not seen as important.

Chapter Eleven

POWER – POLITICAL, FINANCIAL AND ELECTRICAL

Bangladesh government has taken the lead on climate change, playing a key role internationally (Chapter 3) and being among the first developing countries to write action plans. Most money for climate change adaptation has come from government. Donors and lenders have been slow to offer funds, and angry disputes with (and between) donors trying to assert control means some of the offered donor funds were never released – money actually had to be given back to Britain. But the government's leadership position has been tarnished by its own misallocation of funds, by its failure to improve processes of governance (especially in Dhaka and Chittagong) and by the approval of coal-fired power stations which will increase greenhouse gas emissions.

In 2005 the Ministry of Environment and Forest produced the 'National Action Plan on Adaptation' (APA).[1] In 2008 this was transformed into the 'Bangladesh Climate Change Strategy and Action Plan' (BCCSAP) which was the first to be produced by a developing country.[2] In the BCCSAP, Bangladesh opted for a 'pro-poor Climate Change Strategy, which prioritises adaptation and disaster risk reduction.' It has two key aspects. First, it is developmental, stressing the need 'to eradicate poverty and achieve economic and social well-being for all the people'. Second, the plan recognizes that climate change is part of a continuum of climate pressures, and thus stresses the need to 'scale up' existing investments. The Action Plan[3] part of BCCSAP is based on 'six pillars: (1) Food security, social protection and health; (2) Comprehensive disaster management; (3) Infrastructure development; (4) Research and knowledge management; (5) Mitigation and low-carbon development; and (6) Capacity building and institutional development.'[4]

Because climate change amplifies existing climate issues, it is almost impossible to identify specific climate change spending. Are building more shelters and planting trees

---

1 Ministry of Environment and Forests (MEF), 'National Adaptation Programme of Action' (Dhaka: MEF 2005), A. Atiq Rahman, Team Leader. Accessed 13 April 2016, http://unfccc. int/resource/docs/napa/ban01.pdf.This was updated in 2009 as MEF, 'National Adaptation Programme of Action – Updated Version' (Dhaka: MEF 2009), A. Atiq Rahman and Mozaharul Alam, Experts, accessed 13 Apr 2016. http://faolex.fao.org/docs/pdf/bgd149128.pdf.

2 MEF, 'Bangladesh Climate Change Strategy and Action Plan 2008' (Dhaka: MEF, 2008). Accessed 13 April 2016, http://www.adaptation-undp.org/sites/default/files/downloads/ bangladesh_climate_change_actiona_plan.pdf.This was also updated in 2009, accessed 13 April 2016, http://www.climatechangecell.org.bd/Documents/climate_change_strategy2009. pdf.

3 Confusingly, the action plan parts of APA and BCCSAP are different.

4 MEF, 'Bangladesh Climate Change Strategy [...] 2008', xv, 2.

along embankments actions to protect against present cyclones, or actions against the stronger ones expected under climate change? The government spends $1 bn per year – more than 6 per cent of the government budget and 1.1 per cent of GDP – on what the International Centre for Climate Change and Development (ICCCAD) in Dhaka calls 'climate sensitive activities'.[5] Of this three-quarters is domestic resources and one-quarter from foreign donors (of which nearly all are loans, not grants).

In 2010 the government established the Bangladesh Climate Change Trust Fund (BCCTF, or the Trust Fund), and $400 mn was allocated in its first seven years. Some of the money is held back for rapid response to emergencies and disasters. As of June 2015, 297 projects worth $280 mn were being implemented by government and government agencies, while 63 projects worth $3 mn were being implemented by NGOs through the government's Palli Karma-Sahayak Foundation (PKSF). Up to 2013, 45 per cent of Trust Fund money went to the Ministry of Water Resources for dams, rivers and canals, while 14 per cent went to the Forest Department.

With the BCCSAP in 2008, donors began to pledge money. At a climate change conference in London on 10 September 2008, British Prime Minister Gordon Brown pledged £75 mn (then $135 mn) with a second tranche of £75 mn on offer. Donors were unwilling to let the government control the money, so they wanted to set up their own trust fund controlled by the World Bank. The government and civil society were opposed to World Bank control[6] and it was only in 2010 that the Bangladesh Climate Change Resilience Fund (BCCRF or Resilience Fund) was finally established to be managed jointly by the World Bank and the government. It was a disaster. By the end of 2014, seven donors had pledged $181 mn, but only $33 mn had been spent.[7] Of that, more than half ($19 mn) had gone to a World Bank project to build cyclone shelters. The only other two large projects were already government Trust Fund programmes: $8 mn to the Bangladesh Forest Department for tree plantations, and $5 mn to PKSF for NGO projects.

By 2015 it was decided the Resilience Fund would be wound up. Because the World Bank approved so few projects, only a small part of the aid pledged by Gordon Brown was ever used. The second tranche of £75 mn was never released, and it is expected that an unspent £20 mn of the first tranche will be returned to the UK.

Comparisons of government and donor managed funds were obvious. The UK's Independent Commission for Aid Impact (ICAI) pointed out that unlike the Bank man-aged Resilience Fund, the government's own Trust Fund 'got up and running quickly and has invested in developing community resilience'.[8] In 2016, Saleemul Huq, director

5   Md Reaj Morshed et al., 'Good Governance of Climate Change in Bangladesh' (Dhaka: International Centre for Climate Change and Development', 2015) 15.
6   Independent Commission for Aid Impact (ICAI), The Department for International Development's Climate Change Programme in Bangladesh (London: ICAI, 2011).
7   World Bank, BCCRF Annual Report 2014 (Dhaka: Bangladesh Climate Change Resilience Fund [BCCRF] and Ministry of Environment and Forests, 2015), 3–4. Accessed 13 April 2016. https://www.bccrf-bd.org//Documents/pdf/BCCRF%20AR2014_web_resolution.pdf. Note that the report footer incorrectly says 'Annual Report 2013'.
8   ICAI, 1, 14.

of the International Centre for Climate Change and Development in Dhaka said the government's PKSF 'set up a new climate change cell to handle the climate funds it received [...] PKSF is by far the most transparent and, by all accounts, the best run of the various funds'.[9]

Within a year it was already clear that the World Bank management was not working. The UK's ICAI in 2011 reported that the local fears had been justified: 'Concerns focussed on whether the World Bank, rather than the Government of Bangladesh, would be deciding what activities would be funded. DfID's [UK Department for International Development] documentation indicates that this was never the plan, with the role of the World Bank always being that of administrative agent. Evidence from meetings and documents indicates that not everyone within the World Bank originally shared this vision.' The problem was made worse because authority was not 'fully delegated to the World Bank in Bangladesh, with key decisions still being made by staff in its Washington DC headquarters'.[10]

The World Bank is one of the biggest funders of climate change projects in Bangladesh and it asserted its status. The ICAI report noted that DfID was the biggest donor to the Resilience Fund but was 'not holding them [the Bank] sufficiently to account for their performance in implementation'. UNDP is another important climate change player; 'UNDP and the World Bank do not work in a way which is fully collaborative with each other in support of climate change.'[11]

To complaints that the Bank was taking too long to allocate money, the response of one World Bank staff to us was to metaphorically shrug their shoulders and say 'donors wanted quick disbursement, but the World Bank often takes too long for donors because the Bank demands good preparation – up to two years'. Bangladeshis complained that the only way to get funding was to use donor money for existing World Bank projects such as the cyclone shelter building project. Cyclone shelters paid from the Resilience Fund had signs saying the money came from World Bank loans, according to Transparency International Bangladesh (TIB).[12]

All governments have internal struggles, and the battle in Bangladesh over climate change funding was between the ministers of finance and environment – and the donors backed one side. The minister for Environment and Forests was seen as too forceful defending his priorities, while the minister of Finance was seen as more willing to make concessions to donors. With the closure of the Resilience Fund, both the Ministry of Environment and the World Bank retired defeated. Trust funds managed by the World Bank were fashionable with donors in 2010 but had fallen out of fashion by 2015. Instead

---

9 Saleemul Huq, 'Climate Finance in Bangladesh – Learning from Experience', *Daily Star*, 16 April 2016. Accessed 16 April 2016, http://www.thedailystar.net/op-ed/politics/climate-finance-bangladesh-1209307.

10 ICAI, 13–14.

11 ICAI, 1, 14.

12 Mahfuzul Haque, Mohua Rouf and M. Zakir Hossain Khan, *Climate Finance in Bangladesh: Governance Challenges and Way Out* (Dhaka: Transparency International Bangladesh, 2013), 14–15.

donor funds were to go through the Green Climate Fund (GCF), and Bangladesh is among the first group of eight countries to have funds allocated.

The GCF is a stand-alone multilateral financing entity whose sole mandate is to serve the UN Framework Convention on Climate Change (UNFCCC), and thus is accountable to the United Nations. It is governed by a board of 24 members, 12 each from developing and developed countries. Kamal Uddin Ahmed, secretary of the Ministry of Environment and Forests, is a member of the board of the GCF until the end of 2016. But in a victory for the donors, the 'National Designated Authority', which actually deals with the GCF, in not the Ministry of Environment, but the Economic Relations Division of the Ministry of Finance.[13] But the transfer to the Finance Ministry may not work quite the way the donors hope. In a 2015 letter to colleagues, Finance Minister Abul Maal Abdul Muhith said 'we should now direct that Resilience Fund'. In an explicit criticism both of the World Bank and his own government's Trust Fund, Muhith said, 'this fund will not be treated as a new window [for the] water development board or forestry programme'.[14]

Government Corruption and NGO Misconduct

A large number of donor, lender, international NGO and international agency officials pass through Bangladesh. Some take longer than others to learn that Bangladeshis take a stand on technical issues they see as in the national interest, from tidal river management to the choice of open pollinating instead of hybrid rice, and that they have the expertise to take informed decisions. Some consultants never seem to learn. International officials do sometimes feel threatened by Bangladeshis, who are better educated and more knowledgeable than they are. However, Bangladesh's position has been weakened by two issues. The culture of patronage politics and corruption makes it easier for donors to demand control of funds, and the decision to build a large set of coal-fired electricity-generating plants somewhat undermines the government's climate change credentials.

For both donors and local civil society, corruption and political influence are major issues. British policy is clear: 'No UK finance is given directly to the Government of Bangladesh. [...] Instead, DfID uses international organisations and NGOs to manage UK money on its behalf. [...] The widespread nature of corruption in Bangladesh, however, means that continual vigilance is needed.'[15]

Transparency International Bangladesh (TIB) investigated climate change funding and found for one cyclone shelter, site selection, 'project preparation, contractor selection and implementation phase had all been subjected to interference and political influence

---

13 Bangladesh Readiness and Preparatory Support Proposal, 2015. Accessed 14 April 2016, http://www.greenclimate.fund/documents/20182/93876/30122015_-_Bangladesh_Readiness_Proposal.pdf/ca479e02-7e82-4357-842c-e7768ed2fb61.

14 Letter from Finance Minister Abul Maal Abdul Muhit, No-FM/Minister/2015/67, to 16 ministers and secretaries, 1 April 2015.

15 ICAI, 18.

of local influential quarters'. In another shelter, 'sub-contractors had been engaged by the main contractors in complete breach of the procurement rules [...] resulting in the use of low-grade building materials on the part of the sub-contractor'. In some cases projects are supposed to be monitored by a Field Resident Engineer (FRE), but TIB notes that 'FREs are generally in a position of pressure from both LGED [Local Government Engineering Department] and the contractor agencies. In one location, it has been known that the position of FRE has been recruited 12 times within one year as the persons quit the position quickly.' On the other hand, community monitoring can work. TIB cites the case of a shelter and primary school built in Rahmatpur where the school management committee was assigned monitoring responsibility from the outset, and was able to observe construction closely, for example, once discovering substandard stone had been delivered.[16]

The NGO component of the government Trust Fund money was relatively small, but complaints about corruption in the selection of NGOs were published in the press. This was confirmed by TIB, which noted 'exploitation of political power especially by the Ministers, MPs and affiliated government officials'.[17]

NGOs themselves have also been a problem. For example, one NGO, which had a Trust Fund grant to distribute solar panels and carbon-saving stoves, gave them only to its micro-credit borrowers. NGOs received tree-planting grants with no specification of the number of trees or maintenance plans. In four of the projects, non-indigenous species were planted because they were cheaper, but most died because no one took care of the trees.[18]

Meanwhile, relations between NGOs are fraught with tension as they compete for money and often do not collaborate or cooperate. They overlap and repeat projects, often in the places with easiest access. When they are doing relief, different agencies often help the same people, and leave others without assistance because they do not coordinate. Some NGOs replicate others by repairing existing infrastructure, although that is also necessary as NGOs often install pipes or toilets and then move on with no provision for maintenance. In his study of NGOs in slums in Dhaka, Enamul Habib, a senior assistant secretary in the Bangladesh Civil Service, found that 'there is an obvious duplication of beneficiaries and areas covered by different NGOs'.[19] That said, most slum residents will agree that imperfect services from NGOs are better than no services.

International NGOs are an important presence, increasingly as a conduit for foreign donor funds both for their own projects and to fund local NGOs. Habib argues that 'NGO programmes are solely dependent on foreign donors, and this dependency means that the design and guidelines provided by the donors have a significant impact on the way in which NGOs run their health activities. These guidelines often fail to

---

16 Haque, 'Climate Finance', 15–19.
17 Haque, 'Climate Finance', 29.
18 Haque, 'Climate Finance', 37–38.
19 Enamul Habib, 'The Role of Government and NGOs in Slum Development: The Case of Dhaka City', *Development in Practice* 19 (2009), 261.

Table 11.1 Bangladesh's projected emissions reductions in power, transport and industry by 2030

Sector	Base year (2011) ($MtCO_2e$)	BAU scenario (2030) ($MtCO_2e$)	BAU change from 2011 to 2030	Unconditional contribution scenario (2030) ($MtCO_2e$)	Change vs. BAU	Conditional contribution scenario (2030) ($MtCO_2e$)	Change vs. BAU
Power	21	91	336%	86	−5%	75	−15%
Transport	17	37	118%	33	−9%	28	−24%
Industry	26	106	300%	102	−4%	95	−10%
Total	64	234	264%	222	−5%	198	−15%

Note: BAU = Business as Usual.
$MtCO_2e$ = Megatonnes CO_2-equivalent.
Conditional = with 'appropriate international support'.
Source: Table 2, p. 4 of Bangladesh Intended Nationally Determined Contributions, September 2015.

consider socio-cultural conditions and people's perceptions of the slum dwellers.'[20] Ainun Nishat, former vice-chancellor of BRAC University complains that donors and international NGOs come with money but no original thinking – and they won't listen to Bangladeshi experts. A few do interesting pilot projects, but they are small and unsustainable.

Donor Myths and Reality

The often tense relations between Bangladesh and the donors around climate change came up repeatedly in researching this book. 'There was a lot of funding after cyclone Sidr in 2007, but no funding when there are no cyclones. Donors need to see people dying,' commented Mohammad Abdul Quader, a geography professor at Jagannath University. Munir Choudhury, a former joint secretary for Disaster, said, 'donors always want to do relief. How do we push them to fund shelters and other risk reduction that we need?' Donors and the media do not just need disaster, but also climate change disaster, and especially climate change migrants – even though they do not yet exist (see Chapter 8).

Journalists come to Bangladesh wanting stories of climate change disaster, often brought by international NGOs who want to use climate change to raise money. The NGOs cannot say there will be a disaster in 20 years, so journalists and visiting aid officials and diplomats are fed normal environmental disasters as climate change disasters because local experts realize how important the publicity is. The country has some of the world's most eminent climate change experts, but it's also poor and the rich nations assume that wealth gives them wisdom. Bangladeshis are forced to play the game and say precisely what the rich industrialized countries want.

---

20   Habib, 'The Role of Government and NGOs', 261.

Table 11.2 Estimated costs of key adaptation measures

Adaptation measure	Estimated investment required ($bn, 2015–30)
Disaster management	10
Food security and health	8
River flood and erosion	6
Climate resilient infrastructure	5
Salinity intrusion and coastal protection	3
Rural electrification	3
Urban resilience	3
Forestry and ecosystem	2
Wetlands and coastal areas	1
Capacity building	1
Total	42

Source: Table 7, p. 14 of Bangladesh Intended Nationally Determined Contributions, September 2015.

Table 11.3 Estimated costs of key mitigation measures

Mitigation measure	Estimated investment required ($bn, 2015–30)
Supercritical coal power generation	16.5
Solar	3.7
Dhaka metro	2.7
Dhaka expressways	2.6
Increase gas capacity	0.6
Wind	0.6
Biomass from sugar	0.2
Total	26.9

Source: Table 8, p. 14 of Bangladesh Intended Nationally Determined Contributions, September 2015.

Greenhouse Gases

As Bangladesh continues to campaign internationally and put pressure on the already industrialized countries to curb greenhouse gases and to provide money for the loss and damage that is already occurring, the international community also looks at what Bangladesh is promising. As did most other countries, Bangladesh submitted its 'Intended Nationally Determined Contributions' before the Paris COP in 2015.[21] Bangladesh said that its per capita greenhouse gas emissions will not exceed the average for developing countries. It pledges to unconditionally reduce its greenhouse gas emissions by 5 per cent from the 'business as usual' scenario (of no attempt to reduce greenhouse gases – see Chapter 2). Greater reductions are conditional on outside funding, and if the

---

21 Ministry of Environment and Forests (MOEF), 'Intended Nationally Determined Contributions', 2015. Accessed 14 April 2016, http://www4.unfccc.int/submissions/INDC/ Published%20Documents/Bangladesh/1/INDC_2015_of_Bangladesh.pdf.

international community contributes enough money, it will reduce emissions by 15 per cent (see Table 11.1).

But its plans have raised some eyebrows and caused a huge domestic controversy. In April 2016, Bangladesh generated 62 per cent of its electricity from domestic natural gas, 29 per cent from oil and just 2 per cent from coal, with an installed capacity of 12,339 MW.[22] The problem is that gas is running out and Bangladesh is already short of electricity, causing frequent power cuts. So the government's plan is to double electricity production by 2030 with coal-fired power stations, mainly using imported coal.

According to the Intended Nationally Determined Contributions (Table 11.1) the massive investment in highly polluting coal-fired power stations would increase its carbon emissions from electricity generation to more than four times current levels, from 21 $MtCO_2e$ (Megatonnes CO_2-equivalent[23]) to 91 $MtCO_2e$. Bangladesh makes the unconditional promise to reduce this emissions projection by 5 per cent to 86 $MtCO_2e$ but makes the conditional pledge that if donors provided $16.5 bn to fund supercritical coal generation technology (Table 11.3) then this could be reduced to 75 $MtCO_2e$. Coal-fired power stations boil water and the steam runs a turbine that generates the electricity. They have an efficiency of about 32 per cent. Supercritical power plants operate at temperatures and pressures above the critical point of water, at which point there is no difference between water gas and liquid water. This results in higher efficiency of 45 per cent and thus requires less coal, but they are more expensive to build.

The industrialized countries are sufficiently anxious to sell coal technology and are already providing aid and export bank loans for the construction of supercritical coal-fired power stations. Loans and grants are being provided by Japan, which promoted the use of coal in the government's master plan, as well as the United States, India, China and South Korea. Most will be operated by Bangladeshi private companies.

But even with supercritical technology, coal makes a large contribution to greenhouse gases. Because Bangladesh has produced most of its electricity from gas until now, the obvious alternative would be to import liquefied natural gas (LNG). This has clear climate change benefits because natural gas emits about half as much CO_2 as coal.[24] The

---

22 Bangladesh Power Development Board, 'Installed Capacity as on April, 2016, by Fuel Type'. Accessed 14 April 2016, http://www.bpdb.gov.bd/bpdb/index.php?option=com_content&view=article&id=5&Itemid=6.

23 CO_2-equivalent emission is the amount of CO_2 emission that would cause the same warming influence as an emitted amount of a mixture of greenhouse gases, taking into account that some are more persistent in the atmosphere than others.

24 Natural gas is primarily methane (CH_4), which has a higher energy content relative to other fuels, and thus has a relatively lower CO_2-to-energy content. US Energy Information Administration, 'How Much Carbon Dioxide Is Produced When Different Fuels Are Burned?' Accessed 15 April 2016, https://www.eia.gov/tools/faqs/faq.cfm?id=73&t=11. See also Mark Fulton et al., 'Comparing Life-Cycle Greenhouse Gas Emissions from Natural Gas and Coal', Deutsche Bank, 2011.Aaccessed 15 April 2016, http://www.worldwatch.org/system/files/pdf/Natural_Gas_LCA_Update_082511.pdf and A. R. Brandt et al., 'Methane Leaks from North American Natural Gas Systems', *Science* 343 (2014): 735.

Power System Master Plan 2010 is based on a study by the Japan International Cooperation Agency (JICA) and the Tokyo Electric Power Co., which concludes that the 'optimum power supply' is 50 per cent coal. This, in turn, is based on an assumption of high LNG prices and a weighting which gives 70 per cent to economic factors and just 10 per cent to environmental factors.[25] During 2015, LNG prices fell to less than half their former level, bringing it down to the price of coal at the time the plan was being written. Coal prices have also fallen, but not by as much. More supplies of LNG are coming on stream in the early 2020s, so Bangladesh could negotiate cheap long-term supply contracts. Would it make more sense for high income countries to subsidize the price difference between coal and LNG, rather than subsidize polluting coal-fired power plants?

Opting for coal has been heavily criticized in the international press[26] and in Bangladesh. On 4 April 2016, four people were shot dead by police at a protest of thousands of people against a Chinese built and funded coal-fired power plant at Banskhali in Chittagong.[27] On 10–13 March 2016, there was a 250 km protest march from Dhaka to the Sundarbans to protest against the supercritical coal-fired power station under construction at Rampal being built by India's state-owned National Thermal Power Corporation, and another one planned at Khulna. Coal will be imported from Indonesia, South Africa and elsewhere. Boats with coal for Rampal and Khulna will have to pass through the environmentally sensitive Sundarbans. In December 2014 there was a major oil spill when two boats collided inside a dolphin sanctuary in the Sundarbans, which is a mangrove region and UNESCO World Heritage Site.[28]

'As an environmental journalist and as a climate negotiator, I cannot support coal-fired power stations, and I have spoken out publicly,' said Quamrul Chowdhury, a journalist and member of the Bangladesh team at COP talks. 'But the irony, the tragedy, is that I cannot convince my country.'

Conclusion: Contradictory Goals

With its early climate change plans, the government of Bangladesh took an international lead. Most adaptation money has also come from government, which moved rapidly to set up its own Trust Fund and which the UK's Independent Commission for Aid Impact said was 'up and running quickly'. By contrast, there have been ongoing battles with

---

25 Ministry of Power, Energy and Mineral Resources; Japan International Cooperation Agency (JICA) and Tokyo Electric Power Co., *Power System Master Plan 2010*, 8–45. Accessed 15 Apr 2016, http://www.bpdb.gov.bd/download/PSMP/PSMP2010.pdf.

26 Joseph Allchin, 'Bangladesh's Coal Delusion', *The International New York Times*, March 5, 2014, and Joseph Allchin, 'Bangladesh Plans Big Jump in Coal Power', *Financial Times*, 30 December 2015.

27 'Protest at Coal-Fired Power Plant in Banshkali: 4 Villagers Killed in Firing', *New Age*, 5 April 2016. Accessed 15 April 2016, http://newagebd.net/217900/protest-at-coal-fired-power-plant-in-banshkhali/.

28 'Bangladesh Oil Spill "Threatens Rare Dolphins"', *Guardian*, London, 11 December 2014. Accessed 15 April 2016. http://www.theguardian.com/environment/2014/dec/11/bangladesh-oil-spill-threatens-rare-dolphins.

lenders and donor nations, who seem happy to fund coal-fired power stations that will sharply increase greenhouse gas emissions, but less willing to follow Bangladeshi leads on adaptation. Indeed after the World Bank failed to manage a donor Reliance Fund, government has had to hand back unused aid money to the UK.

Government has gained huge credit for its lead in international negotiations and for local leadership on cyclone shelters, early warning systems, rice breeding and so on. But that reputation has been tarnished by corruption and political interference in climate-related projects, which has allowed the donors to justify their reticence at reaching agreement and committing to support Bangladesh's climate change plans.

Saleemul Huq, one of the country's elder statesmen and a highly influential figure on climate change, concluded in 2016: '[P]erhaps most importantly, we need to focus on the need to put in robust systems of transparency of climate fund allocations by governments and other donors, along with robust systems of oversight and accountability by both the sectors of government responsible for those functions, such as the Auditor General and Parliamentary Standing Committees, as well as by third party monitors, including the citizens themselves.'[29]

---

29 Saleemul Huq, 'Climate Finance in Bangladesh [...]', *Daily Star*, 16 April 2016.

Chapter Twelve

BANGLADESH ON THE FRONT LINE OF CLIMATE CHANGE

The Bangladesh Rice Research Institute is developing salt and flood tolerant rice. The national government is building new cyclone shelters. BRAC University architects are designing stronger houses. Climate change is already happening and most studies point to Bangladesh as one of the most vulnerable countries. But it is refusing to be a victim; Bangladesh is not standing by helplessly, waiting for others to act. Instead it is already adapting to climate change and is playing a leading international role in building pressure to halt global warming.

Bangladesh confronts climate change from a basis of knowledge and experience. It has always had an unusual position. It is the world's most densely populated country because it is a rich delta and feeds itself. But that natural wealth has costs in floods and cyclones and a hugely variable climate – which, over centuries, have also made Bangladeshis masters at adaptation. For Bangladesh, climate change does not create new and catastrophic problems – it will not entirely disappear beneath the sea or have huge areas transformed to desert, as will happen in some countries. But nearly one-third of the country is less than 3 m above sea level and the sea is rising; cyclones can kill thousands and all forecasts are that they will get worse; rainfall and flooding will be greater due to global warming. In the country's towns and cities there are likely to be dramatic changes that are not yet understood.

In two different ways, Bangladesh is on the front line of climate change. First, it is highly vulnerable. But second, its history of dealing with an extremely complex and difficult climate made it one of the first to understand the implications of climate change. Climate change was not some theoretical future problem, and Bangladeshis could see what the impact of global warming would be on their country. This led to international campaigning from 1990, and Bangladeshi scientists and intellectuals have been playing a leading role in international negotiations for the past two decades. Inside Bangladesh it led to early consideration of how existing programmes to meet climate problems could be adapted and accelerated to deal with climate change.

In the 45 years since independence, Bangladesh has had remarkable success in dealing with its difficult environment. The worst cyclones used to kill hundreds of thousands, but a post-independence cyclone shelter and early warning programme cut the death rate by 98 per cent. One important response to climate change is building more and stronger shelters which is already underway. At independence, Bangladesh was billed as a famine-ridden 'basket case', but now it feeds itself, partly because of new rice varieties. The shift from monsoon rice production to locally developed irrigated winter *boro* rice took less than a generation and transformed agriculture. So, another important response to climate change is the development of new rice varieties and better agronomic techniques. Faced with waterlogging of the fields, farmers in coastal areas developed ways to use the sediment in the high tides to raise the level of the land. The scientists and engineers named it Tidal River Management and see it as a way of raising land levels to match sea level rise caused by global warming. By using its experience and existing techniques, Bangladesh is already responding to the likely impacts of climate change.

How High Will the Temperature Rise?

For Bangladesh, the Intergovernmental Panel on Climate Change (IPCC) predicts: temperatures will rise, sea level will rise, the number of cyclones will not increase but they will be stronger and more damaging, rainfall and flooding will increase, and while rice production will increase, wheat production will probably become impossible. However the seriousness of these changes depends on the extent of global warming. Bangladesh is pushing for a global agreement to limit maximum temperature increase to 1.5°C over pre-industrial levels, which is 0.5°C over present levels. If that were accepted, the damage done by greenhouse gases would peak in the middle of this century and then decline. The most common proposal is 2°C over pre-industrial levels by the end of this century, which causes increasing damage and rising seas and temperatures well into the next century. Pledges made at the Paris conference in December 2015 would mean a 2.7°C rise in the current century and much more serious damage in the next century.

Greenhouse gases are long-lived in the atmosphere and some of the carbon dioxide emitted now will still be causing climate change a century from now. Bangladesh is pushing for the 1.5°C limit because, in practice, it could cope with that. Adaptation programmes being considered, announced and underway would be sufficient, at least in rural zones. But 2.7°C, the only offer on the table now, will be quite problematic later in this century – for our children and grandchildren – and even worse in the next century.

The Unlivable Cities

Bangladesh's leading role and international standing on climate change adaptation is based on rural areas, where the majority of people still live. Dealing with floods, cyclones, crops and sea level rise has been entirely rural and the expertise and understanding of climate change in Bangladesh by the global experts is built on a deep rural understanding. Unfortunately, Dhaka is a megacity of 18 million people, and lessons from adaptation in rural areas have very limited relevance. It is ranked as one of the least liveable cities in the world, and it becomes ever less able to deal with climate change.

Many factors come into play. The rural areas have a long history and most of the country's intellectuals have rural roots. Political parties look to rural vote banks to win elections. In these areas, patron-client politics is moderated by social networks and a sense that in times of crisis the better off have a social responsibility to the poor. In contrast Dhaka is a new megacity, with a short history and shallow social roots. Its economy is largely informalized. It has grown in the era of neo-liberalism and primacy of the private sector, so the state has played only a limited role in its evolution. What rules exist are easily subverted. The poor majority live in slums and bear the brunt of environmental problems – and climate change; it is their houses that get hottest when the temperature rises; it is their shacks that are flooded when it rains. And they will be evicted as soon as their slum can be replaced with high-rise apartments for the better off.

London at the end of the nineteenth century and Kuala Lumpur, Singapore and Bangkok at the end of the twentieth century made a transition – once they were congested megacities where it was difficult to move and dangerous to breathe, but they became more liveable cities where movement was possible. That transition requires effective urban governance, which in turn requires business leaders to want to change the city and to think long term about prosperity and liveability. Successful cities need a local elite that wants to be proud of their city and wants it to be an acceptable global face of their country. That has not yet happened in Dhaka, where the political and business elite assumes their children will leave Dhaka and Bangladesh and make their lives elsewhere if the city becomes unlivable. Even when the elites do hope their children will return, the children think and do otherwise – why would they come back to the most unlivable city when they can be in the most liveable ones such as London, Melbourne, New York or Toronto.

Climate change will have three main impacts on Dhaka.

- In the flood season rivers are already higher than much of the city; East Dhaka is unprotected and goes under water. Dhaka is caught in the middle by climate change as more water pours down the rivers while sea level rise and storm surges reduce the amount of water that can flow out to the sea, raising river levels near Dhaka. Major embankments are necessary to protect the eastern half of the city and create space for expansion, and existing embankments must be raised.
- Heavy rains flood the part of the city protected by dykes because the water cannot be pumped out quickly enough. Climate change will make this worse because it will bring more intense rainfall. Reopening illegally infilled canals (*khals*) and drainage channels, building storm drains, creating water-holding areas by deepening lakes (*jheels*) and providing adequate pumps will all be needed.
- Cities are normally hotter than the surrounding area and rising temperatures mean Dhaka will suffer from heat stress more often. Many slum-dwellers already report that this is happening. Offices and homes of the better off are air conditioned but the electricity supply already cannot cope. Green infrastructure, improved energy efficiency and housing designs that facilitate air flows will be needed in the future.

Responses to these problems are linked. First, it is the government and government regulated utilities that build infrastructure and provide services and they will need to lead on any response to climate change. We have seen this in rural areas, where it is the national government that leads on rice research, cyclone shelters and other initiatives, but this is not happening in urban areas. Second, there will need to be an acceptance that the poor majority are a part of Dhaka. They have a 'right to the city' and they are central to its economy. They require service provision, including water, sanitation, rubbish collection and electricity. It requires land policy that encourages high-rise, low-cost housing and security of tenure to allow slum dwellers to improve their own homes, to help them cope with higher temperatures and more intense rainfall. And it requires a low-income urban housing strategy. This is a daunting and expensive list, but climate change demands the kind of urban initiatives that are already common in rural areas.[1] Politicians talk of reducing migration, but in the prevailing unregulated practices and markets, rural to urban population shifts are inevitable.

It is difficult for those who know Bangladesh well to imagine such policy reforms but change may have begun. In 2012 Dhaka was split in half, into north and south, and the two mayors elected in 2015 seem more dynamic than past leaders. The current project to save the Hatirjheel, the largest water body inside Dhaka, and turn it into a public open space and holding area for rainwater, shows what is possible – while also showing how influential private interests can still disrupt projects that will deal with climate change. In other parts of the world business people, political leaders and bureaucratic elites have formed coalitions that make cities liveable and make sustained prosperity more probable. One has to hope that similar conditions can develop in Dhaka and Chittagong and look to support them. Such collective action is not purely altruistic as it serves both the self-interest of the elites while meeting the needs of the poor.

The Climate Continuum

For Bangladesh, climate and the response to climate change is a continuum, because climate change makes the normal problems worse. The World Bank estimates that 'by 2050 the total adaptation cost to offset added inundation from climate change is estimated at $5.7 billion' – 5.7 per cent of GDP.[2] The Bank says Bangladesh must invest at least $3.3 bn to protect roads, railways, embankments and other infrastructure from river floods and $2.4 bn in coastal defences and cyclone shelters to respond to climate change. Bangladesh knows how to improve these defences and build the shelters, but that is a daunting amount of money. More money will be needed for Tidal River Management. But if the industrialized and industrializing countries insist on 2.7°C, then the embankments will need to be higher, there will need to be more and stronger cyclone shelters, and Tidal River Management (Chapter 4) will need to be accelerated to keep up with sea level rise.

Similarly, more rain and more floods mean more erosion, more lost land and crops, and more poverty. This, in turn, means more relief and rehabilitation programmes, including cash grants to help millions of people re-establish themselves – or it means

---

1   An alternative would be to try to reduce the role of Dhaka as a primate city which centralizes all political and economic activity. This would mean actively promoting decentralization to smaller cities. Government would need to lead, probably by moving ministries out of Dhaka. Resistance would be substantial, and it may be too late.

2   World Bank, *The Cost of Adapting to Extreme Weather Events in a Changing Climate* (Dhaka: World Bank, 2011), Bangladesh Development Series Paper No. 28, xvi–xvii.

increased poverty which will push more migration to the cities. Cash transfers, often in the form of re-establishment grants for a year or two, will be necessary to keep people in rural areas and allow them to rebuild their livelihoods.

A range of adaptation programmes are underway or possible – high-yielding rice varieties that will withstand longer floods, affordable pumps for water-logged fields, moving houses onto higher plinths and making them stronger to withstand the floods and cyclones and so on. Climate change does not change the problem, just the severity – is a 1 m high house plinth enough, or does global warming require 2 m? Bangladesh knows what is needed but the cost is high – bigger dykes, stronger houses, more pumps and relief payments for those hit by climate change are not free. Bangladesh is not responsible for global warming and the stronger cyclones and rising seas – so who pays for the adaptation? And who decides what is done?

The Bangladesh delta has been spectacularly misunderstood by foreign 'experts'. They ignored the massive watersheds and huge sediment loads and assumed it was like the United States or the Netherlands. Indigenous technical knowledge, local expertise and history were repeatedly dismissed. The rejection of the Flood Action Plan and the introduction of Tidal River Management have been turning points as local knowledge is applied first to reverse damage done by some aid projects, and then to confront climate change.

If the industrialized countries are to pay for the damage they have done – through 'aid', loss and damage payments or green climate funds – will they also insist on saying how the money will be used? Will highly paid foreign consultants continue to be allowed to transfer solutions from other parts of the world and ignore local expertise? The other problem is that aid projects, and particularly climate change projects, have to be presented to rich country taxpayers as 'new' because climate change is seen as something 'new'. Yet in Bangladesh climate change demands more of the same – will donors pay for more cyclone shelters, higher dykes and more relief funds? Or will they say these cannot be seen as climate change adaptation because some already exist? Transforming Dhaka and other cities and defending them against climate change involves massive urban housing construction as well as expanding sewerage and water supply. Will donors realize that these are essential to adapt to climate change, or will they say it is just part of the growth of a megacity and therefore Bangladesh's problem?

As Chapter 11 showed, fund management is likely to be a key issue. So far the international community has tried to keep a tight control over climate change adaptation funds and over the choices as to how they are spent, often ignoring local expertise and experience. Their excuse is usually 'corruption' which is indeed a serious obstacle, but pursuing best practices and achieving nothing may need to be replaced by viable 'second-best' solutions. Part of the problem is that both sides are trying to maintain patron-client systems. The international community wants subservient intermediaries while the Bangladeshi political parties want projects that benefit the influential people in their parties. And both sides have to satisfy people at the top, in capitals or international offices in Geneva or New York, or at senior levels in the party. Which means neither side acts transparently, and each side mistrusts the other. Is this an intractable problem, like Dhaka, or can each side give some ground so that progress can be made?

Keeping Our Heads above Water

Bangladesh can keep its head above water and can maintain the momentum on flood control, response to cyclones, raising coastal land levels and producing more food. Some significant domestic political changes will be essential, especially checking corruption and patronage, and showing the same dynamism and foresight that is already shown in rural areas in the cities too. The governance of Bangladesh is highly imperfect; however, it has delivered steady economic growth for almost 25 years and advanced the country's human development status faster than India.

But Bangladesh is also shaped by outside forces. The delta is built from water and sediment coming from thousands of kilometres away. The cyclones come up the Bay of Bengal. Occupation by the Moghuls, the British and West Pakistan, plus the influence of the cold war and of international agencies, have moulded this land both politically and physically. Sometimes it is essential to adapt to superior forces.

Now global warming is the outside superior force. It is almost entirely created by the wealthy countries from more than a century of fossil fuel emissions, and now pushed by newly industrializing neighbours. The contribution of Bangladesh's rice fields, power stations and garment factories to global warming has been tiny. Yet again, Bangladesh will adapt to outside forces. But many important decisions will be made in the rich, industrialized countries. We see three possible scenarios:

- *Curb emissions:* The most sensible choice for Bangladesh, and for the planet, is a sharp reduction in emissions and meeting the target of limiting global warming to just 1.5°C above pre-industrial levels. Although warming and sea level rise would continue for much of the rest of this century, the damage is manageable and Bangladesh could cope.
- *Emit and pay:* Before the Paris COP 21 meeting, the industrialized, and industrializing nations accepted only a limited curb in emissions, to just 2.7°C at the end of this century. This leads to much more serious damage to the planet and means that Bangladesh would have to make huge adaptation investments – in infrastructure such as embankments, shelters, drains and pumps, and in cash transfers and other support for those whose livelihoods are lost from heavier rain and floods and worse cyclones. At recent COP meetings, the international community has reluctantly accepted the concept of 'loss and damage'. Although not the best choice, if the industrialized countries are prepared to pay for the damage they will continue to cause for the next century, then Bangladesh can adapt.
- *Let the poor pay:* Bangladesh remains one of the poorest countries in the world, but it has used its limited resources relatively wisely to educate its people and raise health and living standards above those of its wealthier neighbour. Will the rich nations say the short-term demands of their own citizen take precedence; that emissions, temperature and sea level must continue to rise; and that they will not pay for the damage they are causing? Will they say that Bangladesh has shown itself skilled in dealing with climate, and must adapt to climate change with its own resources – taking money away from education and social services to raise the dykes and build shelters? Will they say

that those who lose their livelihoods to flood and cyclone should just migrate to the megacity?

Yet again, the future of Bangladesh is partly in the hands of people far away, over whom it has no control and only limited influence. Years of international negotiations in which some of Bangladesh's brightest intellectuals and scientists have played a leading role have brought some gains, with an agreement to prevent runaway global warming and instead limit temperature rise to 2.7°C at the end of the century – which is hugely damaging but at least not catastrophic. And at least lip service has been paid to 'loss and damage'. This is enough to ensure that Bangladesh will keep its head above water – but potentially at a very high cost.

What happens in Bangladesh depends to a significant degree on those of us who live in industrialized countries. Can we push our own governments to lower the temperature rise target to 1.5°C or 2°C? If not, can we at least force them to keep to their 2.7°C pledges? And can we push our governments to pay for the damage we are causing? The rich industrialized countries and their citizens created the global warming that is driving climate change that will raise sea level, increase floods and cause more devastating cyclones. Bangladeshis are already doing more than their fair share and, come what may, will keep their heads above water. Largely using its own money and expertise, Bangladesh is already adapting to climate change. Surely 'we' have a moral duty to help the people of Bangladesh tackle the problems we created.

INDEX